Plants of Central Asia

Volume 14a

Plants of Central Asia

Plant Collections from China and Mongolia

(*Editor-in-Chief*: V.I. Grubov)

Volume 14a
Compositae (Anthemideae)

N.S. Filatova

CRC Press
Taylor & Francis Group
Boca Raton London New York

CRC Press is an imprint of the
Taylor & Francis Group, an **informa** business
A SCIENCE PUBLISHERS BOOK

First published 2007 by Science Publishers Inc.

Published 2019 by CRC Press
Taylor & Francis Group
6000 Broken Sound Parkway NW, Suite 300
Boca Raton, FL 33487-2742

© 2007, Copyright Reserved
CRC Press is an imprint of Taylor & Francis Group, an Informa business

First issued in paperback 2019

No claim to original U.S. Government works

ISBN 13: 978-0-367-45322-0 (pbk)
ISBN 13: 978-1-57808-422-7 (hbk)
ISBN 13: 978-1-57808-062-5 (Set)

Visit the Taylor & Francis Web site at
http://www.taylorandfrancis.com

and the CRC Press Web site at
http://www.crcpress.com

Library of Congress Cataloging-in-Publication Data

Rasteniia Tsentral'noĭ Azii. English
    Plants of Central Asia: plant collections from China and Mongolia
    /[editor-in-chief, V.I. Grubov].
      p. cm.
    Research based on the collections of the V.L. Komarov Botanical Institute.
    Includes bibliographical references.
    Contents: V.14a. Compositae (Anthemideae)
    ISBN 978-1-57808-422-7(vol. 14a)
    1. Botany-Asia, Central. I. Grubov, V.I. II. Botanicheskiĭ institut im. V.L. Komarova. III. Title.
QK374, R23613 2002
581.958-dc21   99-36729
                                                        CIP

ACADEMIA SCIENTIARUM ROSSICA
INSTITUTUM BOTANICUM nomine V.L. KOMAROVII
PLANTAE ASIAE CENTRALIS
(secus materies Instituti botanici nomine V.L. Komarovii)
Fasciculus 14a
COMPOSITAE (ANTHEMIDEAE)
Conficerunt: N.S. Filatova

Translation of: Rasteniya Tsentral'noi Azii, vol. 14a. 2003
                 Izdatel'stvo Sankt-Peterburgskoi gosundarstvennoi
                 khimiko-farmatsevticheskoi akademii (St. Petersburg
                 State Chemical-Pharmaceutical Academy Press),
                 St. Petersburg

ANNOTATION

PLANTS OF CENTRAL ASIA. From the materials of the V.L. Komarov Botanical Institute, Russian Academy of Sciences, vol. 14a: Compositae (Anthemideae). Compiler: N.S. Filatova. 2003. Izdatel'stvo Sankt-Peterburgskoi gosudarstvennoi khimiko-farmatsevticheskoi akademii (State Chemical-Pharmaceutical Academy), St.-Petersburg.

This volume of the illustrated lists of plants of Central Asia (within the People's Republics of China and Mongolia) treats the tribe Anthemideae of the largest family of Compositae. Many members of this tribe, specially wormwoods, tansy, *Brachanthemum* play the most important role in the vegetative cover of Central Asia as coenosis-forming agents (edificators) in steppes and barren lands. This tribe comprises several endemic and relict plants.

As in the preceding volumes, keys are provided for the genera and species and for each species references to nomenclature, its ecology and geographic distribution.

Ill.: 6 plates, 8 maps of distribution ranges.

V.I. Grubov

Editor-in-Chief

PREFACE

3 Tribe Anthemideae treated in this volume occupies a particularly important position in the most numerous family Compositae. In this tribe are concentrated the most important and extensively distributed edificators (characteristic species) forming the coenosis of deserts, barren lands and mountain steppes of Central Asia. Foremost of these are genera *Artemisia* L., *Ajania* Poljak., *Brachanthemum* DC., discussed in detail below. In all, 242 species covering 22 genera are treated in this volume. Adding to them 8 more genera found in former Soviet part of Central Asia, total numbers inhabiting the region rise to 289 species of 30 genera.

In this volume, genus *Artemisia* with 138 species (155 in the entire Central Asia) is the most abundant with respect to the number of species. Rest of genera are incomparably small: *Ajania* 16 species, *Pyrethrum* Zinn 14 species, *Achillea* L. 9 species, and others 1–7 species.

The total number of endemic species in the Central Asian region under consideration here is 36, or 16% of total number of species. Endemic genera here are *Poljakovia* Grub. et Filat. with 3 species and monotypic genus *Stilpnolepis* Krasch. distributed in the southern part of Alashan and Ordos.

In Central Asia as a whole, there are 38 genera of tribe Anthemideae, including 4 endemic genera (*Cancriniella* Tzvel., *Turaniphytum* Poljak. and the 2 genera mentioned before). Similarly, endemic species number 77 of which 36 are wormwood species (20% of total number of species).

As pointed out before, the members of this tribe, along with saltworts, play a leading role in forming the vegetative cover of Central Asia. Particularly characteristic for Central Asia are the associations with species of *Ajania* and *Brachanthemum* DC. predominating. Shrubby *Brachanthemum* species (*B. gobicum* Krasch., *B. mongolicum* Krasch., *B. nanschanicum* Krasch.) represent vegetation defining the landscape of the desert. *B. gobicum* is distributed in the sandy deserts of Eastern Gobi and
4 *B. mongolicum* in the winter fat—saxaul deserts and petrophyte groups of Junggar Gobi while *B. nanschanicum* figures in the saltwort desert associations of Alashan, Qaidam and Qinghai.

Species of *Ajania*, specially desert-steppe species *A. achilleoides* (Turcz.) Poljak. ex Grub., *A. fruticulosa* (Ledeb.) Poljak. and *A. trifida* (Turcz.)

Tzvel. represent important coenosis-forming species of so-called feather grass — wormwood, biurgun — cereal grass, saltwort — wormwood and feather grass — allicaceous desert steppes which cover vast expanses of Mongolia and represent excellent pasture lands while A. *fruticulosa* is extensively distributed in Junggar Gobi as well. A. *roborowskii* Muld. is the main constituent of cereal grass — shrubby — forbs associations of Qaidam, Qinghai and Tibetan mountains.

Genus *Pyrethrum* has no distinctive diversity of species composition. These are mainly mountain species, of which 2 are endemic in Altay while 8 species are distributed in Tien Shan and Pamiro-Alay mountain systems.

Wormwoods occupy a dominant position with respect to the number of species (138 species) as well as the forming of plant associations. In Central Asia, *Artemisia* is represented by 3 subgenera. The largest numbers of species are in subgenus *Artemisia* (69 species) followed by subgenus *Dracunculus* (Bess.) Peterm. (46 species) and subgenus *Seriphidium* (Bess.) Peterm. (23 species). Roughly, a third of the species are associated with Manchurian flora as confirmed by their ranges. Some 10 species are confined to the forest flora (A. *latifolia* Ledeb., A. *sylvatica* Maxim., A. *umbrosa* (Bess.) Pamp., A. *maximovicziana* Krasch. ex Poljak. and others). Most species of this subgenus are, however, mountain xerophytes inhabiting stony, rubbly, rocky slopes, sometimes spread over vast territories. A. *frigida* Willd. represents an edificator and dominant species of arid desert steppes and forms feather grass — wormwood and feather grass — yarrow — wormwood communities on rubbly trails of mountains and conical hillocks of entire Central Asia. A. *xerophytica* Krasch., A. *rutifolia* Steph. ex Spreng. and A. *santolinifolia* (Turcz. ex Pamp.) Krasch. represent typical petrophytes of rocky and rubbly slopes figuring in the shrubby desert-steppe associations in all the Central Asian regions.

Wormwoods of subgenus *Dracunculus* are also associated with Manchurian flora (A. *manshurica* (Kom.) Kom., A. *eriopoda* Bge.) but a predominant number of species form an element of Gobi flora. In the desert regions, specially in the sandy deserts, they form landscape plants comprising vegetative coenosis over vast expanses and represent the basic nutritious fodder in autumn-winter pastures. A. *xanthochroa* Krasch. is distributed in Eastern Mongolia, Gobi Altay, Junggar and Eastern Gobi and Alashan in hummocky and sand dunes, in saxaul forests, rarely on rubbly-sandy trails; A. *klementzae* Krasch. ex Leonova in semi-fixed, barhan and thin sand of Mongolian (including lake basins) and Gobi Altay; A. *xylorhiza* Krasch. ex Filat. on hummocky and semifixed sand, rarely on sandstone outcrops of Eastern Mongolia, Gobi Altay, Eastern Gobi and Alashan; A. *ordosica* Krasch. on semifixed and barhan sand of

Gobi Altay, Alashan, Ordos, Khesi and northern part of Qaidam; *A. sphaerocephala* Krasch. on thin, rubbly and ridge sand, sometimes in solonchak lowlands and depressions between sand dunes in Gobi Altay, Eastern, Western, Junggar Gobi and Alashan.

5   *A. duthreuil-de-rhinsii* Krasch., *A. pewzowii* Winkl. and *A. nanschanica* Krasch. figure in the high-mountain barren steppes of Kunlun, Qinghai and Tibet; *A. borealis* L., *A. depauperata* Krasch. and others in the high mountains of Mongolia and Gobi Altay.

Subgenus *Seriphidium* is not known for the abundance of species in this territory. Its distribution range is restricted mainly to the western regions of Central Asia, predominantly Junggar and Tien Shan but this subgenus occupies the leading position with respect to the diversity of species and formation of vegetative associations. *A. sublessingiana* Krasch. ex Poljak., an eastern Kazakhstan species, is dominant in the steppified wormwood—feather grass associations of north-western Junggar Gobi and foothills of the lower belt of southern slope of Altay. *A. gorjaevii* Poljak. is dominant in the wormwood—cereal grass mountain steppes (at 1500–2000 m altitudes) of Junggar-Tarbagatai and Northern Tien Shan. *A. heptapotamica* Poljak. represents the main constituent of ephemeral wormwood, often with forbs, barren steppes of Junggar Gobi, Junggar and Northern Tien Shan. *A. subchrysolepis* Filat. is a dominant species in hammada (rocky) type desert, forms wormwood—saltwort and wormwood with saxaul associations in rubbly and rubbly-rocky deserts of Junggar-Tarbagatai and Northern Tien Shan. *A. issykkulensis* Poljak. forms wormwood—bean caper, wormwood—Reaumuria coenoses at altitudes of up to 2000 m in Central Tien Shan and Kashgar. Northern Turan species *A. terrae-albae* Krasch., dominant in the desert zone, is reported in Tarbagatai and Junggar Gobi. *A. gracilescens* Krasch. et Iljin is a regional desert-steppe species forming shrubby cereal grass—wormwood associations on solonetzic, rubbly-rocky slopes of low mountains and trails of Junggar Ala Tau and Junggar Gobi. *A. schischkinii* Krasch. is a typical member of mountain desert steppes with Nanophyton on the southern slopes of Mongolian Altay and Junggar Gobi. *A. elongata* Filat. et Ladyg. dominates midmountain wormwood-cereal grass with pea shrub steppes (1600–2600 m) of Kashgar, Junggar and Northern Tien Shan. *A. borotalensis* Poljak. is dominant in the high-mountain wormwood—cereal grass steppes (2200–3200 m) of Mongolian Altay, Junggar Ala Tau and Northern Tien Shan. *A. skorniakovii* Winkl. is dominant in the high-mountain deserts (3600–4000 m) of Kashgar and Pamir.

The following are typical halophytes of Central Tien Shan: *A. saissanica* (Krasch.) Filat and *A. schrenkiana* Ledeb. in meadowy and puffed solonchaks of Junggar Gobi; *A. nitrosa* Web. ex Stehm. in solonetzic meadows, along banks of saline lakes of Eastern Mongolia; *A. gobica* (Krasch.) Grub. in solonetzic meadows, columnar solonetzes, Gobi clays of Eastern Mongolia, Valleys of Lakes, Gobi Altay, Eastern and Western Gobi, Alashan; *A. assurgens* Filat. in the marshy rocky-clayey soils, solonetzic meadows, saline sand of Gobi Altay, Junggar Gobi and Alashan.

The above report reveals that members of Anthemideae play an extremely significant role in the flora and vegetation of Central Asia. An overwhelming number of species are landscape and characteristic (edificators) species of plant associations.

In conclusion, we would like to emphasize once again the extreme diversity of genus *Poljakovia*. While describing his new species *Tanacetum falcatolobatum* Krasch., I.M. Krascheninnikov drew attention to the absence so far of unanimity on genus *Tanacetum*. According to him, in south-eastern Mongolia and adjoining Chinese province, original species and (as confirmed by our own data) genera from the morphological viewpoint could be found. Genus *Poljakovia* is a xerophilised member of upland xerophytes found in the mountains of south-eastern Mongolia and adjoining mountain regions of China.

Monotypic genus *Stilpnolepis* inhabits the deserts of Alashan and Ordos exclusively on sand. It was originally described by K. Maximowicz as *Artemisia centriflora* Maxim. but differs greatly due to a variety of factors, morphological as well as genetic, from *Artemisia*. A highly characteristic feature of genus *Stilpnolepis* is the phyllary in the form of broad orbicular scales that are broad-scarious along margins, semi-transparent, and highly lustrous in spite of faint pubescence. It is for this very characteristic that the genus has acquired its name.

These 2 original genera, as well as some members of other genera, eloquently confirm the speciation processes in this territory.

H   H   H

For this volume, O.I. Starikova translated the Chinese references and herbarium labels. Artist O.V. Zaitseva prepared the drawings for plates while N.S. Filatova plotted the maps of distribution ranges and compiled the indexes.

V.I. Grubov and N.S. Filatova

CONTENTS

TAXONOMY

SPECIAL ABBREVIATIONS

Abbreviations of Names of Collectors

Bar.	— V.I. Baranov
Beket.	— U. Beket
Chaff.	— J. Chaffanjon
Chaney	— R.W. Chaney
Ching	— R.C. Ching
Chu	— C.N. Chu
Czet.	— S.S. Czetyrkin
Dar.	— Sh. Dariima
Divn.	— D.A. Divnogorskaya
Fed.	— B.A. Fedtschenko (Fedczenko)
Fet.	— A.M. Fetisov
Glag.	— S.A. Glagolev
Golubk.	— N.S. Golubkova
Gr.-Grzh.	— G.E. Grum-Grzhimailo
Grombch.	— B.L. Grombchevski
Grub.	— V.I. Grubov
Gub.	— I.A. Gubanov
Gus.	— V.A. Gusev
Ik.-Gal.	— N.P. Ikonnikov-Galitzkij
Isach.	— E.A. Isachenko
Ivan.	— A.F. Ivanov
Kal.	— A.V. Kalinina
Kam.	— R.V. Kamelin
Karam.	— Z.V. Karamysheva
Klem.	— E.N. Klements
Kom.	— V.L. Komarov
Krasch.	— I.M. Krascheninnikov

Kryl.	—	P.N. Krylov
Kuan	—	K.C. Kuan
Lad.	—	V.F. Ladygin
Ladyzh.	—	M.V. Ladyzhensky
Lavr.	—	E.M. Lavrenko
Lee, Lee et Chu, and also Lee et al	—	A.R. Lee (1959)
S.H. Li	—	S.H. Li et al (1951)
Lipsk.	—	V.I. Lipsky
Lis.	—	V.I. Lisovsky
Litw.	—	D.I. Litwinow
Lom.	—	A.M. Lomonossov
Merzb.	—	G. Merzbacher
Mois.	—	V.S. Moiseenko
Muld.	—	A.A. Muldashev
Pal.	—	I.V. Palibin
Pavl.	—	N.V. Pavlov
Petr.	—	M.P. Petrov
Pewz.	—	M.V. Pewzov
Pias.	—	P.Ya. Piassezki
Pob.	—	E.G. Pobedimova
Pop.	—	M.G. Popov
Pot.	—	G.N. Potanin
Przew.	—	N.M. Przewalsky
Rachk.	—	E.I. Rachkovskaya
A. Reg.	—	A. Regel
Rhins	—	J.L. Dutreuil de Rhins
Rob.	—	V.I. Roborowsky
Sap.	—	V.V. Sapozhnikov
Schischk.	—	B.K. Schischkin
Serp.	—	V.I. Serpukhov
Sold.	—	V.V. Soldatov
Tug.	—	A.Ya. Tugarinov
Ulzij.	—	N. Ulzijkhutag
Volk.	—	E.A. Volkova
Wang	—	K.S. Wang
Yun.	—	A.A. Yunatov

| Zab. | — D.K. Zabolotnyi |
| Zam. | — B.M. Zamatkinov |

Abbreviations of Names of Herbaria

B	— Botanischer Garten und Botanisches Museum, Berlin-Dahlem
BM	— The Natural History Museum, London
DUH	— Botany Department, University of Delhi, New Delhi
E	— Royal Botanic Garden, Edinburgh
Fl	— Herbarium Universitatis Florentinae, Museo Botanico, Firenze
G	— Conservatoire et Jardin botaniques de la Ville de Geneve, Geneve
HIMC	— Biology Department, University of Inner Mongolia, Hohhot
K	— Royal Botanic Gardens, Kew, London
KW	— M.G. Kholodny Institute of Botany, The Ukrainian Academy of Sciences, Kiev
L	— Rijksherbarium, Leiden
Linn.	— The Linnean Society of London, London
M	— Botanische Staatsammlung, München
MW	— Herbarium of the Moscow State University, Moscow
P	— Museum National d'Histoire Naturelle, Paris
PE	— Herbarium, Institute of Botany, Academia Sinica, Beijing
PR	— National Museum in Prague, Průhonice
S	— Swedish Museum of Natural History, Stockholm
TAK	— Tashkent State University, Tashkent
TI	— Botanical Gardens and University Museum, University of Tokyo, Tokyo
TK	— Krylov Herbarium of Tomsk State University, Tomsk
UPS	— Botanical Museum, Uppsala University, Uppsala
W	— Department of Botany, Naturhistorisches Museum, Wien (Vienna)
WRSL	— Museum of Natural History, Wroclaw University, Wroclaw

Family 118. COMPOSITAE Giseke
Subfamily CARDUOIDEAE Cass. ex Sweet
Tribe Anthemideae Cass.

1. Thalamus invariably covered with well-developed scaly scarious bracts .... 2.

+ Thalamus glabrous or faintly pilose but without scaly scarious bracts (or the latter seen only on ray flowers in anthodium) .... 4.

2. Anthodium homogamous, all flowers in it tubular, bisexual; monocarp with compact white tomentose pubescence .... 3. Handelia Heimerl.

+ Anthodium heterogamous; ray flowers in it pistillate, ligulate, disc (central) flowers bisexual, tubular .... 3.

3. Annual plants, scabrous due to sparse pubescence or glabrous; anthodium single, large; ray flowers pistillate, barren; thalamus convex .... 1. Anthemis L.

+ Perennial plants, invariably pubescent with long hairs; anthodia small, form loose or compact corymbose inflorescence; ray flowers pistillate, fertile; thalamus flat or subconvex .... 2. Achillea L.

4. Anthodium homogamous, with many bisexual tubular disc flowers; ligulate pistillate ray flowers absent in anthodium .... 5.

+ Anthodia heterogamous, many or single, invariably with pistillate ray or tubular flowers .... 7.

5. Perennial plants with thick lignifying multicipital rhizome, with rather few fertile and vegetative shoots, sometimes somewhat lignifying at base; anthodia form compact or loose corymb .... 16. Hippolytia Poljak.

+ Subshrubs or tiny subshrubs with many fertile and vegetative shoots forming compact or loose mats; anthodia form panicle or single (sometimes 2–3) .... 6.

6. Anthodia single, aggregated on terminal upper part of branching shoots; phyllary coriaceous, faintly pubescent .... 11. Poljakovia Grub. et Filat.

+ Anthodia form compact or loose panicle; phyllary herbaceous, soft, with compact tomentose pubescence .... 22. Artemisia L. (subgenus Seriphidium (Bess.) Peterm.).

7. Ray flowers in anthodium fertile, part pistillate, part bisexual; innermost flowers and those sessile at tip of thalamus, staminate, barren; annual plant with cristate-laciniate leaves .... 19. Neopallasia Poljak.

+   Plant with different set of characteristics .... 8.

8.   Achenes with 2–3 highly exserting longitudinal nerves, prismatic, with 2 red resinous glandules at tip; annual-biennial plants .... 6. Tripleurospermum Sch.-Bip.

+   Achenes with 5–10 (rarely 4) barely exserting, often poorly noticeable longitudinal nerves .... 9.

9.   Thalamus convex, flat or subglobose; achenes with pappus..10.

+   Thalamus convex but achenes invariably without pappus or with barely noticeable fringe at tip .... 14.

10.   Thalamus convex, conical, punctate-tuberculate, glabrous; anthodium on long leafless pedicel, single, small, 7–10 mm in diam., or anthodia even form a corymb and then plant annual .... 11.

+   Thalamus slightly or highly convex, globose; anthodium, if single, 10–20 mm in diam.; if, however, in a corymb, small, up to 10 mm in diam. .... 12.

11.   Anthodia form loose corymb; phyllary green, lustrous, diffuse-pilose, subobtuse or rarely short-acuminate at tip; achenes up to 1.2 mm long .... 5. Matricaria L.

+   Anthodium single, many on plant; phyllary with whitish tomentose pubescence, cuspidate at tip; achenes 1.2–1.6 mm long .... 9. Cancrinia Kar. et Kir.

12.   Anthodia invariably form compact or loose corymbose inflorescence; ray flowers in anthodium yellow or whitish yellow .... 10. Tanacetum L.

+   Anthodium single or 2–8 together but do not form corymbose inflorescence; ray flowers in anthodium white, pink or some other shade, but not yellow .... 13.

13.   Perennial herbaceous plant with erect fertile shoots, 15–35 (40) cm tall; thalamus more or less punctate-glandular, sometimes with noticeable cellularity .... 7. Pyrethrum L.

+   Perennial herbaceous plant up to 15 cm tall, with highly branched fertile and vegetative shoots at base forming loose, scattered mat; thalamus very weakly punctate-tuberculate .... 8. Waldheimia Kar. et Kir.

14 (9).   Ray flowers in anthodium ligulate, disc (central) flowers tubular .... 15.

+   All flowers in anthodium tubular .... 16.

15.   Anthodium 4–7 mm in diam.; ray flowers 1–10, their corolla yellowish white with broad-ovoid or broad-oblong short (1–8 mm long) limb .... 13. Brachanthemum DC.

+   Anthodium 8–20 (30) mm in diam.; ray flowers 10–30, their corolla white, pink or some other shade (but not yellow), with oblong or linear, 10–30 mm long limb .... 12. Chrysanthemum L.

16.   Anthodia invariably form compact or loose panicle; achenes tiny, obovoid or oblong, somewhat flattened .... 22. Artemisia L.

+   Anthodia several, form inflorescence of different type or single; achenes of different type, invariably laterally flattened .... 17.

17.   Subshrubs, tiny subshrubs or perennial herbaceous plants ... 18.

+   Annual plants with easily pulled out long root and many large single anthodia .... 22.

18.   Monocarp with profuse cobwebby-tomentose pubescence and thick erect stem (sometimes, stems 2–3); anthodia many, form compact umbelliform compound corymb .... 4. Pseudohandelia Tzvel.

+   Plants pubescent or altogether glabrous; anthodia form simple corymbose inflorescence, sometimes single .... 19.

19.   Herbaceous perennial plants, covered at base of fertile shoots with fibrous stalk remnants of year-old leaves .... 20.

+   Subshrubs or tiny subshrubs with shoots lignifying at base .... 21.

20.   Fertile shoots altogether glabrous; leaf blade profile obovoid, 2–3-pinnatisected; terminal leaf lobes filiform-linear, up to 4 cm long, with short cusp at tip; anthodia short-stalked, form corymb .... 17. Filifolium Kitam.

+   Fertile shoots pubescent with long cobwebby hairs; leaf blade narrow-lanceolate, entire, tomentose-pubescent underneath; anthodia sessile, form compact corymb .... 14. Phaeostigma Muld.

21.   Subshrub with entire leaves or latter 3–5-lobed at tip, pubescent with sparse stellate hairs; anthodia gathered in bundles at branch ends .... 18. Kaschgaria Poljak.

+   Tiny subshrub with more complexly laciniated leaf blades, pubescent with simple or double-ended hairs; anthodia form corymb, rarely single .... 15. Ajania Poljak.

22 (17).   Annual plant with narrow crescent-shaped simple leaves; phyllary scarious, scale-like, glabrous .... 20. Stilpnolepis Krasch.

+   Annual plant with tripartite leaves; phyllary pubescent .... 21. Elachanthemum Ling et Y.R. Ling.

1. Anthemis L.

Sp. pl. 2 (1753) 893

1. **A. cotula** L. Sp. pl. 2 (1773) 894; Fedtsch. in Fl. SSSR, 26 (1961) 63; Fl. Sin. 76, 1 (1983) 8; Fl. Intramong. ed. 2, 4 (1993) 559; Opred. rast. Sr. Azii [Key to Plants of Mid. Asia] 10 (1993) 519. — *Matricaria cotula* (L.) DC. Prodr. 6 (1838) 13. — Ic.: Fl. SSSR, l.c. Plate 4, fig. 1; Fl. Sin. l.c. tab. 3, fig. 1; Fl. Intramong. l.c. tab. 221, fig. 1.

Described from Europe (Ukraine). Type in London (Linn.).

On debris, roadsides, around houses, in plantations.

IA. Mongolia: *East. Mong.* (Huh-Hoto-Fl. Sin. l.c., Fl. Intramong. l.c.).

General distribution: Europe (except the north), Caucasus, Mid. Asia.

Note. Weed. Fl. Sin. cites this species for East. Mongolia. Evidently introduced. This species has been included in our treatment since plants depicted in the cited drawings correspond wholly to the diagnosis and the find of this species in Mongolian territory is entirely likely.

2. Achillea L.

Sp. pl. (1753) 896

1. Leaves entire, serrato-dentate along margin .... 1. A. acuminata (Ledeb.) Sch.-Bip.

+ Leaves pinnatilobed or pinnatisected .... 2.

2. Blades of lower cauline leaves narrow-linear, with tiny (up to 1–2 (3) mm long) transversely imbricate segments .... 5.

+ Blade of lower cauline leaves more broad, with much larger (up to 3–5 mm long), usually upward, not transverse nor imbricate, segments .... 3.

3. Stems usually single or 2–3 together, faintly pubescent with slender short white hairs; phyllary green, keeled, with exserted midnerve, with brownish scarious border .... 6. A. millefolium L.

+ Stems rather few, greyish or greyish green due to long, slender, tangled hairs; phyllary yellowish-greenish, weakly keeled, with brown or whitish scarious border .... 4.

4. Leaf blade 2–3-pinnatisected; leaf lobes linear or linear-lanceolate, bristly; anthodia in compact, dense, subcapitate corymbs; ray flowers yellowish white, whitish underneath .... 8. A. setacea Waldst. et Kit.

+ Leaf blade bipinnatisected; leaf lobes oblong or oblong-linear, not bristly; anthodia in loose corymbs; ray flowers pink or red, rarely white .... 3. A. asiatica Serg.

5. Rhizome long; leaf blade pinnatisected for not more than up to centre with entire lanceolate lobes; phyllary with brown membranous border .... 5. A. ledebourii Heimerl.

+ Rhizome short; leaf blade pinnatisected or pinnatilobed up to axial nerve with linear or linear-lanceolate lobes that are unevenly dentate or pinnate along margin; phyllary with much lighter-coloured membranous border .... 6.

6. Leaf blade pinnatisected, its lowermost segments with 1–2 large teeth on each side apart from fine teeth; upper segments with different-sized fine teeth, rarely without teeth; ray flowers white with violet base or rarely violet .... 4. A. impatiens L.

+ Leaf blade pinnatilobed or pinnatipartite with linear or linear-lanceolate lobes, in turn, pinnatipartite; terminal lobes serrato-dentate along margin; ray flowers white .... 7.

7. Leaf blade pinnatisected, with profuse punctate glandules; anthodia oblong, many; phyllary with yellowish or narrow brown border; flowers covered with capitate glandules ....
.... 7. A. ptarmicoides Maxim.

+ Leaf blade pinnatilobed or pinnatipartite, without punctate glandules or with a small number of them; anthodia semiovoid, rather few; phyllary yellowish, with much broader brown border, without capitate glandules .... 2. A. alpina L.

1. A. acuminata (Ledeb.) Sch.-Bip. in Flora, 38 (1855) 15; Boczantzev in Fl. SSSR, 26 (1961) 108; Grub. Opred. rast. Mong. [Key to Plants of Mongolia] (1982) 243; Fl. Sin. 76, 1 (1983) 13; Fl. Intramong. ed. 2, 4 (1993) 560; Gub. Konsp. fl. Vneshn. Mong. [Conspectus of Flora of Outer Mongolia] (1996) 95.

Described from East. Siberia (Transbaikal). Type in St.-Petersburg (LE). Plate I, fig. 1.

In meadows, scrubs, floodplains and along river banks.

IA. Mongolia: *East. Mong.* (Kuku-Khoton, meadow around Begin-nor lake, Sept. 2, 1884 – Pot.; "Ulantsab" – Fl. Intramong. l.c.), *Ordos* (Alashan mountains, Huang He river valley, July 28, 1871 – Przew.).

General distribution: East. Sib., Far East, Nor. Mong. (Hent., Mong.-Daur., Cis-Hing.), China (Dunbei), Korean peninsula, Japan.

2. A. alpina L. Sp. pl. (1753) 899; Boczantzev in Fl. SSSR, 26 (1961) 119; Grub. Opred. rast. Mong. [Key to Plants of Mongolia] (1982) 242; Fl. Sin. 76, 1 (1983) 16; Fl. Intramong. ed. 2, 4 (1993) 562; Gub. Konsp. fl. Vneshn. Mong. [Conspectus of Flora of Outer Mongolia] (1996) 95. – *A. mongolica* Fisch. ex Spreng. Nov. proven. hort. Acad. Halen. et Berol. (1818) 3; Heimerl in Denkschr. Acad. Wiss. Math. – Nat. Wien, 48, 2 (1884) 127, in clave. – *Ptarmica alpina* (L.) DC. Prodr. 6 (1838) 22, p.p. quoad pl. sibir. –

13 *P. mongolica* (Fisch. ex Spreng.) DC. Prodr. 6 (1838) 22. — *P. sibirica* Ledeb. Fl. Ross. 2, 2 (1845) 528; Kom. et Alis. Opred. rast. Dal'nevost. kraya [Key to Plants of Far East] 2 (1932) 1027; Kitam. Compos. Japon. 2 (1940) 324. — Ic.: Grub. l.c. Plate 89, fig. 590.

Described from Siberia. Type in London (Linn.).

In forest and forest-steppe belts of mountains, along river banks, coastal pebble beds, in meadows, willow thickets, scrubs.

IA. Mongolia: *East. Mong.* (Dariganga, Shilin-Bogdo-ula, extinct volcano, nor. slope, cereal grass — forbs meadow, Aug. 11, 1970 — Grub., Ulzij., Tserenbalzhid; valley of khalkhin-gol river, 20 km above Sumber settlement, June 25, 1987 — Kam., Gub., Dar. et al; low mountains westward of road between Batnar and Narovlin somons, 35 km from latter, willow thickets in upper part of nor. slope, Aug. 11, 1989-Grub., Gub. et Dar.; "Shilin-Khoto"-Fl. Intramong. l.c.).

General distribution: East. Sib., Far East, Nor. Mong., China (Dunbei).

3. A. asiatica Serg. in Animadv. syst. Herb. Univ. Tomsk, 1 (1946) 6; Kryl. Fl. Zap. Sib. 11 (1949) 2723; Grub. in Not. syst. (Leningrad) 19 (1959) 551; Afan. in Fl. SSSR, 26 (1961) 85; Fl. Kirgiz. 11 (1965) 111; Fl. Kazakhst. 9 (1966) 12; Grub. Opred. rast. Mong. [Key to Plants of Mongolia] (1982) 243; Fl. Sin. 76, 1 (1983) 12; Fl. Intramong. ed. 2, 4 (1993) 560; Opred. rast. Sr. Azii [Key to Plants of Mid. Asia] 10 (1993) 523; Gub. Konsp. fl. Vneshn. Mong. [Conspectus of Flora of Outer Mongolia] (1996) 95.

Described from East. Siberia (Chulym river valley). Type in Tomsk (TK).

In larch forests and along their borders, in flooded and dry-valley meadows, scrubs.

IA. Mongolia: *Mong. Alt.* (Kara-tyr (Chernyi Kobdo), cape in lake vicinity, Aug. 1, 1908 — Sap.; valley of Urtugol river, larch forest along nor. slope of mountains, Aug. 17, 1930-Pob.; 30 km south of Tamchi-Daba pass, midcourse of Bidzhigol, near spring, Aug. 10; same site, west. Bulgan somon, on road to Khara-Khatu-Khutul', on lower boundary of forest, Aug. 27-1947, Yun.; south. extremity of Dayan-nur lake, nor. slope of Yamatyn-ul toward valley, along lower fringe of forest, 2350 m, sedge — alkali grass meadow, July 10, 1971 — Grub., Ulzij. et Dar.; valley of Ulyastyn-gol, Shadzgat-nuru mountain range, 2100 m, *Cobresia* meadow, June 27, 1973 — Golubk. et Tsogt; mountains along south. bank of Dayan-nur lake, floor of creek valley, marshy coastal meadow, July 26, 1977 — Karam. et Sanczir; basin of Uenchiin-gol, valley of Arshantiin-gol (right tributary of Khargaityn-gol), 3 km from estuary, in larch forest, at 2500-2600 m alt., Aug. 14, 1979-Grub., Muld. et Dar.; basin of Bulgan river, upper reaches of Ulyastyin-gol gorge, left bank tributaries, July 9, 1984 — Kam. et Dar.), *Cent. Khalkha* (basin of Dzhargalanta river, upper reaches of Uber-Dzhargalanta river, near Botoga mountains, aspen forest, mountain-meadow belt, Sept. 2, 1925 — Krasch. et Zam.), *East. Mong.* (Kulun-Buirnor plain, nor.-west. slope of Malagaiten-Daban pass, July 11, 1899 — Pot. et Sold.; Yakeshi station, arid mountain slope, July 9, 1954 — K.-Ch. Wang).

IIA. Junggar: *Tien Shan* (Merzbacher mountain range, valley of Erdynkho river, high erosion terrace, at 1100-1200 m, July 9, 1952-Mois.).

General distribution: Fore Balkh., Jung.-Tarb., West. and East. Sib., Far East, Nor. Mong., China (Dunbei), Japan.

4. A. impatiens L. Sp. pl. (1753) 858; Heimerl in Denkschr. Acad. Wiss. Math.-Natur. Wien, 48 (1884) 127, in clave; Boczantzev in Fl. SSSR, 26 (1961) 116; Fl. Sin. 76, 1 (1983) 14; Gub. Konsp. fl. Vneshn. Mong. [Conspectus of Flora of Outer Mongolia] (1996) 95. — *Ptarmica impatiens* (L.) DC. Prodr. 6 (1838) 22; Serg. — in Fl. Zap. Sib. 11 (1949) 2731. — Ic.: Fl. Sin. tab. 3, fig. 2.

Described from Siberia. Type in London (Linn.).

In mountain meadows, forest and mountain meadows.

14   IA. Mongolia: *Cent. Khalkha* (upper course of Uber-Dzhargalanta river, near Botoga mountains, aspen forest, mountain meadow, Sept. 2, 1925 — Krasch. et Zam.).

General distribution: West. and East. Sib., Nor. Mong. (Hent., Hang., Mong.-Daur.).

5. A. ledebourii Heimerl in Flora (1883) 389 and in Denkschr. Acad. Wiss. Math.-Natur. Wien, 48, 2 (1884) 127, in clave; Boczantzev in Fl. SSSR, 26 (1961) 117; Grub. Opred. rast. Mong. [Key to Plants of Mongolia] (1982) 243; Fl. Sin. 76, 1 (1983) 14; Gub. Konsp. fl. Venshn. Mong. [Conspectus of Flora of Outer Mongolia] (1996) 95. — *Ptarmica krylovii* Serg. in Animadv. syst. Herb. Univ. Tomsk, 1–2 (1949) 13 and in Fl. Zap. Sib. 11 (1949) 2730. — *P. tenuisecta* (Kryl.) Serg. in Animadv. syst. Herb. Univ. Tomsk, 1–2 (1949) 13 and in Fl. Zap. Sib. 11 (1949) 2731. — *A. alpina* auct. non L.: Ledeb. Fl. Ross. 2, 2 (1845) 528.

Described from Altay. Type in Wien (Vienna) (W).

In alpine and subalpine meadows.

IA. Mongolia: *Cent. Khalkha* (40 km nor.-east of Sumber somon, Bayan-Berkheng-obo mountain, forbs — cereal grass steppe, Aug. 8, 1970 — Mirkin).

General distribution: West. Sib. (Altay).

6. A. millefolium L. Sp. pl. (1753) 899; Kryl. Fl. Zap. Sib. 11 (1949) 2721; Grub. Konsp. Fl. MNR [Conspectus of Flora of Mongolian People's Republic] (1955) 243; Afan. in Fl. SSSR, 26 (1961) 78; Fl. Kazakhst. 9 (1966) 10; Grub. Opred. rast. Mong. [Key to Plants of Mongolia] (1982) 243; Fl. Sin. 76, 1 (1983) 10; Fl. Intramong. ed. 2, 4 (1993) 560; Opred. rast. Sr. Azii [Key to Plants of Mid. Asia] 10 (1993) 522. — *A. millefolium* var. *vulgaris* Trautv. in Bull. Soc. natur. Moscou, 39, 1 (1866) 345. — *A. setacea* auct. non Waldst. et Kit.: Fl. Kirgiz. 11 (1965) 110, p. max. p. — *A. asiatica* auct. non Serg.; Fl. Kirgiz. l.c. 111; Fl. Kazakhst. 9 (1966) 12, p.p.

Described from West. Europe. Type in London (Linn.).

In forest and forest-steppe zones, forest borders, meadows, along river banks, around houses, along roadsides, farm borders, in tugais.

IIA. Junggar: *Cis-Alt.* (Oi-Chilik, Sept. 8, 1876-Pot.), *Zaisan* (between Kara area and village on Kaba river, on bank of irrigation ditch, June 16; same site, around Kaba village, tugai, June 16, 1914—Schischk.), *Tien Shan* (Borgaty river, 1874—Lad.; Talki river gorge, July 18; Sairam, above 1000 m, July 20; Kul'd-zha, July–1877, A. Reg.; Talki river gorge, 1500 m, July 10; Sairam, July 23; Shara-Bugutal pass, Aug.–1878, Fet.; Dzhisumtas (Kash river), 1000 m, July 2; Mongoto (Kash river), about 2000 m, Aug. 12–1879, A. Reg.; valley of Bol. Yuldus, Aug. 5; Bogdo-ula mountains, on nor,. slope of pass, Aug. 30–1898, Klem.; Bogdo-ula and Urumchi, steppe slopes, Aug. 2–3; vicinity of Ulan-Ussu, Aug. 15–17—1908, Merzb.).

General distribution: Aralo-Casp., Fore Balkh., Jung.-Tarb., Nor. Tien Shan, Europe, Mid. Asia (West. Tien Shan), Nor. Mong. (Hent., Hang., Mong.-Daur.).

7. **A. ptarmicoides** Maxim. Prim. Fl. Amur. (1859) 154; Heimerl in Denkschr. Acad. Wiss. Math.-Natur. Wien, 48 (1884) 127, in clave; Kitag. Lin. Fl. Mansh. (1939) 241; Boczantzev in Fl. SSSR, 26 (1961) 121; Grub. Opred. rast. Mong. [Key to Plants of Mongolia] (1982) 243; Fl. Sin. 76, 1 (1983) 18; Fl. Intramong. ed. 2, 4 (1993) 561. —Ic.: Fl. Intramong. l.c. tab. 222, fig. 1–4.

Described from Far East. (Amur river). Type in St.-Petersburg (LE). Plate I, fig. 2.

In forbs—cereal grass valley and mountain, rarely solonetzic meadows, scrubs, uremas (bottomland deciduous forests).

IA. Mongolia: *East. Mong.* (sand dunes around Bagin-nor lake, Oct. 21, 1884—Pot.), "Baotou"-Fl. Intramong. l.c.), *Cent. Khalkha* (floodplain of Kerulen river, east of Bayan-Berkheng-obo mountain, solonetzic floodplain meadow, Aug. 8, 1970—Mirkin et Kashapov), *East. Mong.* (valley of Khalkhin-gol river, 13 km south-east of Khamar-Daban, forbs—cereal grass meadow in floodplain, Aug. 11, 1949—Yun.; "Khailar, Shilin-Khoto, Yakeshi, Khukh-Khoto"—Fl. Intramong. l.c.).

General distribution: East. Sib., Far East, China (Dunbei), Japan.

8. **A. setacea** Waldst. et Kit. Pl. rar. Hung. 1 (1802) 82; Kryl. Fl. Zap. Sib. 11 (1949) 2725; Afan. in Fl. SSSR, 26 (1961) 83; Fl. Kirgiz. 11 (1965) 110, p. min. p.; Fl. Kazakhst. 9 (1966) 11; Fl. Sin. 76, 1 (1983) 11; Opred. rast. Sr. Azii [Key to Plants of Mid. Asia] 10 (1993) 522.—*A. millefolium* L. var. *setacea* (Waldst. et Kit.) Koch, Syn. Fl. Germ. ed. 1 (1837) 373; Ledeb. Fl. Ross. 2, 6 (1845) 532, p.p. —Ic.: Fl. Sin. l.c. tab. 3, fig. 3.

Described from West Europe (Hungary). Type in Wien (Vienna) (W).

In steppes, forest borders, scrubs, rarely on meadow and meadow-steppe slopes, as weed in plantations, roadsides, on fallow land and pastures.

IIA. Junggar: *Cis-Alt.* (20 km nor.-west of Shara-Sume, shrubby meadow-steppe, Aug. 7; same site, 105 km nor.-nor.-west of Ertai settlement on road to Kok-Togai, one of the sources of Chern. Irtysh, meadow on sandy alluvium, July 14–1959, Yun. et Yuan').

General distribution: Aralo-Casp., Fore Balkh., Jung.-Tarb., Nor. and Cent. Tien Shan, Europe, Caucasus, West. Sib.

3. Handelia Heimerl
in Österr. Bot. Zeitschr. 70 (1922) 215

1. H. trichophylla (Schrenk) Heimerl in Österr. Bot. Zeitschr. 70 (1922) 215; Tsvetk. in Fl. SSSR, 26 (1961) 125; Fl. Kirgiz. 11 (1965) 113; Fl. Kazakhst. 9 (1966) 17; Fl. Sin. 76, 1 (1983) 19; Opred. rast. Sr. Azii [Key to Plants of Mid. Asia] 10 (1993) 525. — Achillea trichophylla Schrenk in Fisch. et Mey. Enum. Pl. Nov. 1 (1841) 48; Ledeb. Fl. Ross. 2, 6 (1845) 538. — Ic.: Fl. Kazakhst. l.c. Plate 2, fig. 7.

Described from East. Kazakhstan (Junggar Ala Tau). Type in St.-Petersburg (LE).

On rubbly and melkozem slopes of low mountains, along arid river beds, as weed in plantations and in fallow lands.

IIA. Junggar: Dzhark. (Piluchi, near Kul'dzha, June 20, 1877-A. Reg.), Tien Shan (Talkibash, July 19, 1877 — A. Reg.; "Nor. Sinkiang (entire region) — Fl. Sin. l.c.).

General distribution: Jung.-Tarb, West. Tien Shan, Nor. Tien Shan, Mid. Asia (plains and mountains, Pam. — Alay).

4. Pseudohandelia Tzvel.
in Fl. URSS, 26 (1961) 878

1. P. umbellifera (Boiss.) Tzvel. in Fl. URSS, 26 (1961) 878, 363; Fl. Kazakhst. 9 (1966) 67; Opred. rast. Sr. Azii [Key to Plants of Mid. Asia] 10 (1993) 527. — Tanacetum umbelliferum Boiss. Diagn. Pl. Or. Nov. ser. 2, 3 (1856) 30. — T. trichophyllum Rgl. et Schmalh. in Acta Horti Petrop. 5, 2 (1877) 255. — Ic.: Fl. SSSR, l.c. Plate 15; Fl. Kazakhst. l.c. Plate 8, fig. 5.

Described from Afghanistan. Type in Geneva (G). Isotype in St.-Petersburg (LE).

On rubbly and fine-sandy slopes of low mountains, limestone outcrops, on sand, as weed on fallow lands.

IIA. Junggar: Dzhark. (Kul'dzha, June 13, 30, 1877 — A. Reg.).

16   General distribution: Fore Balkh., Jung.-Tarb., Nor. Tien Shan, Mid. Asia (West. Tien Shan, Pam.-Alay), Fore Asia (Iran, Afghanistan).

5. Matricaria L.
Sp. pl. (1753) 1256

1. Leaf blade 1–2-pinnatisected, terminal leaf lobes narrow-linear; anthodia heterogamous, on long peduncles, ray flowers white, disc flowers yellow, 5-toothed .... 2. M. recutita L.

+ Leaf blade bipinnatisected, terminal leaf lobes linear; anthodia homogamous (without marginal flowers), on much shorter peduncles, disc flowers greenish yellow, 4-toothed ....
.... 1. M. matricarioides (Less.) Porter ex Britton.

1. M. matricarioides (Less.) Porter ex Britton in Mem. Torrey Bot. Club, 5 (1884) 341; Kitam. in Mem. Coll. Sci. Kyoto Imp. Univ. ser. B. (Comp. Jap.), 15 (1940) 334; Pobed. in Fl. SSSR, 26 (1961) 150; Fl. Kirgiz. 11 (1965) 116; Fl. Kazakhst. 9 (1966) 20; Fl. Sin. 76, 1 (1983) 50; Fl. Intramong. ed. 2, 4 (1993) 574. — *Artemisia matricarioides* Less. in Linnaea, 6 (1831) 210, excl. syn. — *Matricaria discoidea* DC. Prodr. 6 (1838) 50; Ledeb. Fl. Ross. 2, 6 (1845) 544. — *Lepidotheca suaveolens* (Pursh) Nutt. in Transact. Amer. Philos. Soc. 7 (1841) 397; Opred. rast. Sr. Azii [Key to Plants of Mid. Asia] 10 (1993) 620. — Ic.: Fl. Kazakhst. l.c. Plate 2, fig. 6; Fl. Sin. l.c. tab. 6, fig. 2; Fl. Intramong. l.c. tab. 227, fig. 1–3.

Described from Unalashka island. Type in St.-Petersburg (LE).

Weed on roadsides, around residences, in gardens, pastures.

IA. Mongolia: *East. Mong.* ("Yakeshi town" — Fl. Sin. l.c., Fl. Intramong. l.c.).

General distribution: Europe, West. and East. Sib., Far East, Nor. Amer.

2. M. recutita L. Sp. pl. (1753) 891; Pobed. in Fl. SSSR, 26 (1961) 148; Fl. Kirgiz. 11 (1965) 116; Fl. Kazakhst. 9 (1966) 20; Grub. Opred. rast. Mong. [Key to Plants of Mongolia] (1982) 243; Gub. Konsp. fl. Vneshn. Mong. [Conspectus of Flora of Outer Mongolia] (1996) 103. — *M. chamomilla* L. Sp. pl. ed. 2 (1763) 1256; Ledeb. Fl. Alt. 4 (1833) 114; Ledeb. Fl. Ross. 2, 6 (1845) 545. — *Chamomilla recutita* (L.) Rauschert, Fol. Geobot. Phytotax. 9, 3 (1974) 255; Opred. rast. Sr. Azii [Key to Plants of Mid. Asia] 10 (1993) 620. — Ic.: Fl. Kazakhst. l.c. Plate 2, fig. 5; Fl. Sin. l.c. tab. 6, fig. 1.

Described from West. Europe. Type in London (Linn.).

Weed in fields, fallow lands, around roads, residences, in gardens.

IIA. Junggar: *Jung. Gobi* ("Nor.-West. Sinkiang" — Fl. Sin. l.c.).

General distribution: Nor. Tien Shan, Europe, West. and East. Sib., Far East, Nor. Mong. (Hent.), Nor. Amer.

6. Tripleurospermum Sch.-Bip.
Tanacet. (1844) 32

1. Perennial plant with woody decumbent rhizome bearing some fertile shoots and leaf rosettes at base; phyllary with dark brown undulate, membranous border ....
.... 1. T. ambiguum (Ledeb.) Franch. et Sav.

+ Annual-biennial plant with one or more easily pulled out fertile shoots, without leaf rosettes at base; phyllary with light-coloured margin .... 2. T. inodorum (L.) Sch.-Bip.

1. T. ambiguum (Ledeb.) Franch. et Sav. Enum. 1 (1875) 236, quoad nomen; Pobed. in Fl. SSSR, 26 (1961) 168; Fl. Kazakhst. 9 (1966) 22; Fl. Sin. 76, 1 (1983) 51; Gub. Konsp. fl. Vneshn. Mong. [Conspectus of Flora of Outer Mongolia] (1996) 109. — *Pyrethrum ambiguum* Ledeb. Fl. Alt. 4 (1833) 118; Ledeb. Fl. Ross. 2, 6 (1845) 547. — *Matricaria ambigua* Kryl. Fl. Alt. 3 (1904) 625; Opred. rast. Sr. Azii [Key to Plants of Mid. Asia] 10 (1993) 618. — Ic.: Fl. SSSR, l.c. Plate 7, fig. 1; Fl. Kazakhst. l.c. Plate 2, fig. 1.

Described from Altay. Type in St.-Petersburg (LE). Map 1.

In alpine and subalpine meadows, larch forests, along banks of rivers, 2000–3500 m alt.

IA. Mongolia: *Mong. Alt.* (westward of Bulgan somon on road to Khar Tantu-Khutul', at lower forest boundary, July 27, 1947 — Yun.; Kobdo river basin, west, slope of Tsengel-Khairkhan mountain range, upper course of Khargantu-gol river, on road to Dayan-nur, nor. slope of Yamatyn-ul, larch forest at 2350–2500 m alt., July 9, 1971 — Grub., Ulzij., Dar.; mountains on southern bank of Dayan-nur lake, meadow along bank, July 26, 1977-Karam, Sanczir, Sumerina).

IIA. Junggar: *Cis-Alt.* (Oi-Chilik, alp. belt, Aug. 20; Kengeity river, tributary of Chern. Irtysh, Sept. 18–1876, Pot.; Kran river, Aug. 24, 1903 — Gr. Grzh.; upper course of Ustyugan river, tributary of Kran river, alpine meadow, July 1, 1908 — Sap.), *Tien Shan* (Kakkamyr mountains, July; Koktyube, July 23; Bogdo mountains, 2700–3000 m, July 24, 1878; between Kash river and Nilki, July 8; Karagol, 3000 m, July 14 and 17–1879, A. Reg.; Sairam, July 12, 1878 — Fet.; under Mukhurdai pass, 3500–3800 m, July 19, 1893 — Rob.; in Kunges region, on slope in shade, No. 1561, Aug. 7, 1958 — A.R. Lee), Jung. Alat. (westward of Ven'tsyuan' village, in waterdivide, No. 2018, Aug. 5; mountain in Toli region, nor. slope of waterdivide, No. 1746, Aug. 6–1958 — Shen).

General distribution: Jung.-Tarb., Nor. Tien Shan, West. Sib.

2. T. inodorum (L.) Sch.-Bip. Tanacet. (1844) 32; Pobed. in Fl. SSSR, 26 (1961) 175; Fi. Kirgiz, 11 (1965) 119; Fl. Kazakhst. 9 (1966) 23; Fl. Sin. 76, 1 (1983) 53. — *Matricaria inodora* L. Fl. Suec. ed. 2 (1755) 765; DC. Prodr. 6 (1838) 52; Ledeb. Fl. Ross. 2 (1845) 545; Kryl. Fl. Zap. Sib. 11 (1949) 2733; Opred. rast. Sr. Azii [Key to Plants of Mid. Asia] 10 (1993) 618. — Ic.: Fl. Kazakhst. l.c. Plate 2, fig. 2; Fl. Sin. l.c. tab. 6, fig. 3.

Described from Sweden. Type in London (Linn.).

Along banks of rivers, lakes, as weed in plantations, on roadsides.

IIA. Junggar: *Tarb.* ("Dachen" - Fl. Sin. l.c.), *Tien Shan* (Sairam, July 12, 1878 — Fet.; between Karakol and Nilki, June 16, 1879; Baibeshan, Aug. 31, 1880 — A. Reg.; Nilki, on slopes, No. 3993 — Kuan).

General distribution: Aralo-Casp., Fore Balkh., Nor. Tien Shan, Europe, Caucasus, Mid. Asia (West. Tien Shan, Pam.-Alay).

Note. The find of *T. homogamum* G.X. Fu in nor. Sinkiang (Burchum river) cited in Fl. Sin. l.c. has not been substantiated in our plant collections.

7. Pyrethrum Zinn
Catal. Pl. Gütting. (1757) 414;

Scop. Fl. Carn. ed. 2, 2 (1772) 148

1. Plants 25–75 cm tall, with one or more fertile shoots; radical and lower cauline leaves on long (not shorter than blade) petioles, green, glabrous or diffuse-pilose, 2–3-pinnatisected .... 2.

+ Plants 5–35 cm tall, generally form carpet-like mats along with shortened vegetative shoots; radical and lower cauline leaves on much shorter petioles, compactly pubescent, 1–2-pinnatisected .... 9.

2. Anthodia up to 15–18 mm in diam., generally 2–8 of them singly on long pedicels emerging from axils of cauline leaves; do not form a regular corymb .... 3.

+ Anthodia up to 14–25 mm in diam., singly on long pedicels emerging from axils of only upper cauline leaves .... 7. P. krylovianum Krasch.

3. Fertile shoots rather few, somewhat foliate; lower cauline leaves without perceptible glandules; terminal leaf lobules filiform-linear to linear-lanceolate, long cuspidate at tip .... 2. P. alatavicum (Herd.) O. et B. Fedtsch.

+ Fertile shoots single, rarely 2–3 together, uniformly foliate; lower cauline leaves punctate-glandular; terminal leaf lobules oblong, short-cuspidate at tip .... 4.

4. Fertile shoots glabrous, foveolate-punctate-glandular; peduncles only under anthodia with loose pubescence of long flexuose hairs; phyllary with board dark-brown margin .... 5.

+ Fertile shoots with loose woolly pubescence; peduncles pubescent all along their length with long flexuose hairs; phyllary with very narrow dark brown margin .... 6.

5. Fertile shoots single, rarely 2, uniformly foliate; lower cauline leaves green, bipinnatisected, glandular; terminal leaf lobules oblong, thickish; short-cuspidate at tip .... 4. P. changaicum Krasch. ex Grub.

+ Fertile shoots few, somewhat foliate; lower cauline leaves tripinnatisected, without perceptible glandular pubescence; terminal leaf lobules linear to filiform-linear, soft, long-cuspidate at tip .... 1. P. abrotanifolium Bge. ex Ledeb.

6. Radical and lower cauline leaves usually dull greyish-green due to loose tomentose pubescence; terminal leaf lobules lanceolate,

short-cuspidate at tip, spaced ....
.... 8. P. lanuginosum (Sch.-Bip. et Herd.) Tzvel.

+   Radical and lower cauline leaves green, rarely diffuse-pilose, subglabrous; terminal leaf lobules narrow-linear or linear-lanceolate, long-cuspidate at tip .... 7.

7.   Limb of corolla of ray flowers linear or narrow-linear; corolla of tubular flowers 3–3.5 mm long .... 9. P. pulchrum Ledeb.

+   Limb of corolla of ray flowers broad-linear; corolla of tubular flowers 2–3 mm long .... 8.

8.   Lower and middle cauline leaves petiolate with winged margin at base, invariably without tomentum in axils; terminal leaf lobules linear or linear-lanceolate; anthodium single, on long loose pubescent pedicels .... 5. P. karelinii Krasch.

+   Lower and middle cauline leaves on very short petioles without winged margin at base, without compact tomentum in axils; terminal leaf lobules broad-lanceolate; anthodia 2–3, at ends of fertile shoots .... 11. P. richterioides (Winkl.) Krassn.

9 (1).   Plants with branched rather thin rhizome, with many generative and vegetative shoots forming small but compact mats; anthodium up to 10 (11) mm in diam. .... 10.

+   Plant with thick multicipital rhizome, with many vegetative and stray or 2–3 generative shoots, forming quite large, loose mats; anthodium up to 15 mm in diam. .... 12.

10.   Fertile shoots subglabrous or very diffusely pilose; lower and middle cauline leaves altogether glabrous; green, leaf blade oblong in profile, without punctate glandules; anthodium on long peduncles without pubescence; phyllary glabrous ....
.... 3. P. arassanicum (Winkl.) O. et B. Fedtsch.

+   Fertile shoots with greyish pubescence; lower and middle cauline leaves greyish green due to fairly compact tomentose pubescence of long hairs; leaf blade profile oblong-linear, with punctate glandules; anthodium on pedicels with tomentose pubescence; phyllary (specially at base) with tomentose pubescence .... 11.

11.   Leaf blade of lower and middle cauline leaves bipinnatisected; terminal leaf lobules lanceolate to lanceolate-ovoid, with short chondroid cusp at tip ....
.... 10. P. pyrethroides (Kar. et Kir.) B. Fedtsch. ex Krasch.

+   Leaf blade of lower and middle cauline leaves tripinnatisected; terminal leaf lobules lanceolate to lanceolate-filiform, with long chondroid cusp at tip ....
.... 14. P. transiliense (Herd.) Rgl. et Schmalh.

12. Radical and lower cauline leaves greyish green due to tomentose pubescence of long hairs; leaf blade bipinnatisected; terminal leaf lobules linear to linear-lanceolate; crown of achene split into oblong glumes up to base .... 13.

+ Radical and lower cauline leaves whitish-greyish due to compact tomentose pubescence; leaf blade tripinnatisected; terminal leaf lobules lanceolate or lanceolate-ovoid; crown of achene entire or dentate along margin ....
.... 13. P. songoricum Tzvel.

13. Entire plant with whitish tomentose pubescence; fertile shoots many, ascending; leaves whitish tomentose, short-stalked; leaf blade profile oblong, up to 4 cm long; terminal leaf lobules oblong, short-cuspidate at tip; phyllary coriaceous-herbaceous .... 12. P. semenovii (Herd.) Winkl. ex. O. et B. Fedtsch.

20 + Plant with greyish tomentose pubescence; fertile shoots rather few (3–5), erect; leaves greyish-tomentose, long-stalked; leaf blade profile oblong-ovoid, up to 8 cm long; terminal leaf lobules lanceolate, subulate-cuspidate at tip; phyllary soft, herbaceous .... 6. P. kaschgaricum Krasch.

1. P. abrotanifolium Bge. ex Ledeb. Fl. Ross. 2, 2 (1845) 549; Kryl. Fl. Zap. Sib. 11 (1949) 2767; Tzvel. in Fl. SSSR, 26 (1961) 250; Fl. Kazakhst. 9 (1996) 34; Fl. Sin. 76, 1 (1983) 68; Opred. rast. Sr. Azii [Key to Plants of Mid. Asia] 10 (1993) 607. — *Chrysanthemum abrotanifolium* (Bge.) Kryl. Fl. Alt. 3 (1904) 621, p.p. — Ic.: Fl. Sin. l.c. tab. 9, fig. 3.

Described from Altay. Type in St.-Petersburg (LE).

On rocky and rubbly slopes, 2000–3000 m alt., rarely on forest borders.

IB. Kashgar: *Nor.* ("Tarim basin" — Fl. Sin. l.c.).

IIA. Junggar: *Tien Shan* (3 km south of Yakou, nor. slope, No. 1668 and 1714, Aug. 30, 1957 — Shen; 60 km nor. of Ulyastai, Nilke district, on slope, No. 3993, Aug. 31, 1957 — Kuan; Kunges, Narat mountain range, on shaded slope, 2300 m, No. 6554, Aug. 7, 1958 — A.R. Lee (1959)).

General distribution: Jung.-Tarb.; West. Sib. (Altay).

2. P. alatavicum (Herd.) O. et B. Fedtsch. Konsp. Fl. Turkest. [Conspectus of Flora of Turkestan] 4 (1911) 186; Kryl. Fl. Zap. Sib. 11 (1949) 2748; Grub. in Not. syst. (Leningrad), 19 (1959) 352; Tzvel. in Fl. SSSR, 26 (1961) 249; Fl. Kirgiz. 11 (1965) 123; Fl. Kazakhst. 9 (1966) 32; Grub. Opred. rast. Mong [Key to Plants of Mongolia] (1982) 244; Fl. Sin. 76, 1 (1983) 66; Opred. rast. Sr. Azii [Key to Plants of Mid Asia] 10 (1993) 606; Gub. Konsp. fl. Vneshn. Mong. [Conspectus of Flora of Outer Mongolia] (1996) 104. — *Tanacetum alatavicum* Herd. in Bull. Soc. natur. Moscou, 40, 3 (1867) 129. — *Chrysanthemum alatavicum* (Herd.) B. Fedtsch. Rast. Turkest. [Plants of Turkestan] (1915) 737. — Ic.: Fl. Kazakhst. l.c. Plate 3, fig. 1.

Described from East. Kazakhstan (Trans-Ili Ala Tau). Type in St.-Petersburg (LE).

On rocky and rubbly slopes, along banks of rivers, above 2500 m alt.

IA. Mongolia: *Mong. Alt.* (Fl. Sin. l.c.; Gub. l.c.).

IIA. Junggar: *Tien Shan* (Talkibash, around 3000 m, July 22, 1877; Sairam, around 3000 m, Aug. 2; Kokkamyr mountains, Aug. 27, 1878; Aryslyn estuary, around 3000 m, July 20, 1879—A. Reg.; upper course of Tekes river, meadow, Aug. 24, 1903—Rob.).

General distribution: Jung.-Tarb., Nor. Tien Shan, Mid. Asia (West. Tien Shan, Pam.-Alay).

3. **P. arassanicum** (Winkl.) O. et B. Fedtsch. in Konsp. Fl. Turk. [Conspectus of Flora of Turkestan] 4 (1911) 187; Tzvel. in Fl. SSSR, 26 (1961) 259; Fl. Sin. 76, 1 (1983) 70.— *Chrysanthemum arassanicum* Winkl. in Acta Horti Petrop. 11, 12 (1891) 372.— *Pyrethrum pyrethroides* auct. non (Kar. et Kir.) B. Fedtsch. ex Krasch.; Fl. Kirgiz. 11 (1965) 125; Opred. rast. Sr. Azii [Key to Plants of Mid. Asia] 10 (1993) 608. —Ic.: Fl. Sin. l.c.: tab. 10, fig. 3.

Described from Kirghizia (Kirghiz mountain range). Type in St.-Petersburg (LE).

On rocky and rubbly slopes, rock talus at 2500–4000 m alt.

IIA. Junggar: *Tien Shan* (Uch-Turfan, Karagaily gorge, on bed of brook among stones, June 17; same site, Airy gorge, on rock talus, June 19—1908, Divn.).

IIIC. Pamir (Fl. Sin. l.c.).

General distribution: Nor. and Cent. Tien Shan, Mid. Asia (West. Tien Shan, Pam.-Alay).

4. **P. changaicum** Krasch. ex Grub. in Not. syst. (Leningrad), 17 (1955) 23; Grub. in Opred. rast. Mong [Key to Plants of Mongolia] (1982) 244; Gub. Konsp. fl. Vneshn. Mong. [Conspectus of Flora of Outer Mongolia] (1996) 104. —Ic.: Grub.; l.c. Plate 80, fig. 591.

21 Described from Mongolia (Hangay). Type in St.-Petersburg (LE).

In larch forests and their borders, along brooks, among rocks, boulders and on rocks, at upper forest boundary.

IA. Mongolia: *Mong. Alt.* (Gub. Konsp. l.c.).

General distribution: Nor. Mong.; endemic.

5. **P. karelinii** Krasch. in Not. syst. (Leningrad), 9 (1946) 157; Tzvel. in Fl. SSSR, 26 (1961) 254; Fl. Kirgiz. 11 (1965) 124; Fl. Kazakhst. 9 (1966) 35; Opred. rast. Sr. Azii [Key to Plants of Mid. Asia] 10 (1993) 607.— *Chrysanthemum richterioides* Winkl. var. *virescens* Winkl. in Acta Horti Petrop. 10 (1887) 86.— *Tripleurospermum pulchrum* (Ledeb.) Rupr. in Mem. Acad. Sci. Petersb. 7, 14 (1869) 52, quoad. pl.— *Tanacetum pulchrum* auct. non (Ledeb.) Sch.-Bip.; Herder in Bull. Soc. natur. Moscou, 40, 2 (1867) 129.— *Pyrethrum pulchrum* auct. non Ledeb.: Fl. Kazakhst. 9 (1966) 34, p. max. p. —Ic.: Fl. Kazakhst. l.c. Plate 3, fig. 2.

Described from East. Kazakhstan (Junggar Ala Tau). Type in St.-Petersburg (LE).

On rocky slopes of high mountains, subalpine grasslands, in mountain tundras at 2500–5000 m alt.

IIA. Junggar: *Tien Shan* (Talkibash-Sairam, July 12; Kash river, July 18; valley of Murzat river, 3300 m, Aug. 18; Kunges river, Aug. 22 - 1877, A. Reg.; Dzhagastai-gol, July 19; Aryslan, 3000 m; Monguto, Aug. 4–1879, A. Reg.; Urten-Muzart, Aug. 3, 1877 – Fet.; around Mukhurdai pass, 3500–3700 m, July 19, 1893 – Rob.; Koksu, 3400 m, Oct. 8, 1907 – Merzb.), *Jung. Alat.* (20 km south of Ven'tsyuan' town, No. 3763, Aug. 14, 1957 – A.R. Lee (1959)).

IIIC. Pamir (in upper courses of Kanlyk river, 4500–5000 m, mountain tundra, July 4; sources of Kashka-su river, 4200–5500 m, July 5–1942, Serp.).

General distribution: Jung.-Tarb., Nor. and Cent. Tien Shan, Mid. Asia (West. Tien Shan, Pam.-Alay).

6. P. kaschgaricum Krasch. in Not. syst. (Leningrad), 9 (1946) 158; Fl. Sin. 76, 1 (1983) 71. —Ic.: Fl. Sin. l.c. tab. 8, fig. 3.

Described from Sinkiang (East. Tien Shan). Type in St.-Petersburg (LE). Plate II, fig. 3.

On arid steppe slopes of mountains, 2000-2600 m alt.

IB. Kashgar: *West.* (west. Kun-Lun, King-tau mountain range, 40–50 km south-west of Upal settlement, steppe belt of mountains, 2600 m alt., June 10, 1959 – Yun.).

IIA. Junggar: *Tien Shan* (East. Tien Shan (Biangou area, Sept. 25, 1929 – Pop., typus !).

General distribution: endemic in Cent. Asia.

7. P. krylovianum Krasch. in Not. syst. (Leningrad), 9 (1946) 155; Kryl. Fl. Zap. Sib. 11 (1949) 2746; Tzvel. in Fl. SSSR, 26 (1961) 250; Fl. Sin. 76, 1 (1983) 66; Gub. Konsp. fl. Vneshn. Mong. [Conspectus of Flora of Outer Mongolia] (1996) 104. — *Chrysanthemum abrotanifolium* (Bge.) Kryl. in Fl. Alt. 3 (1904) 621, p.p. — *Ch. merzbacheri* B. Fedtsch. ex G. Merzbacher, Gebirgs-gruppe Bogdo-Ola (1916) 312. —Ic.: Fl. Sin. l.c. tab. 9, fig. 2.

Described from Altay. Type in St.-Petersburg (LE).

On rocky slopes, rocks of upper belt of mountains.

IA. Mongolia: *Mong. Alt.* (Fl. Sin. l.c.; Gub. l.c.).

IIA. Junggar: *Cis-Alt.* (Shara-Sume, along fringe of ditch, No. 1956 and 3763, Aug. 6, 1956 – Ching), *Tien Shan* (Mai-Tyube pass, Talki mountain range, Sept. 4, 1953 – Mois.; Kul'dzha, 1878, No. 298 – Lar.; at foot of Bogdo-ul, July 1878; Kash river, July 15; estuary of Aryslan river, July 8–1879, A. Reg.; Bogdo-Ola and opp. Urumtschi, Nach. Fukan in getrocknen Lehm- und Kiessteppe massenhaft, No. 1276, Aug. 2, 1908 – Merzb.).

General distribution: West. Sib. (Altay).

Note. A comparison of types of *P. krylovianum* Krasch. and *P. merzbacheri* B. Fedtsch. did not reveal any differences between them apart from the colour of ray flowers. A study of herbarium material and their geographic habitats also led to a similar conclusion. Fl. Sin. (l.c.) cited *P. merzbacheri* B. Fedtsch. among synonyms of *P.*

richterioides (Winkl.) Krasch., based only on the colour of ray flowers although *P. merzbacheri* B. Fedtsch. does not differ at all from *P. krylovianum* Krasch. in the nature of pubescence, leaf blade division, and many other characteristics but the persistence of the pink colour of ray flowers needs to be confirmed.

8. **P. lanuginosum** (Sch.-Bip. et Herd.) Tzvel. in Fl. SSSR, 26 (1961) 252; Grub. Opred. rast. Mong. [Key to Plants of Mongolia] (1982) 243; Gub. Konsp. fl. Vneshn. Mong. [Conspectus of Flora of Outer Mongolia] (1996) 104. — *Tanacetum lanuginosum* Sch.-Bip. et Herd. in Pollichia, 20–21 (1863) 442.

Described from East. Sib. (East Sayans, Munku-Sardyk mountain). Type in St.-Petersburg (LE).

On rocky and fine-granular slopes in high mountains, among rocks, on turf-covered talus and placers.

IA. Mongolia: *Khobd.* (south-east. offshoot of Turgen mountain range, Kashgut mountain, 3200 m, among rocks, July 21, 1977 — Karam., Sanczir, Sumerina; Tsagan-Shibetu mountains at Ureg-nur, 3000 m, on summit, Aug. 4, 1978 — Karam., Beket. et al.), *Mong. Alt.* (pass from Mal. Kairta area in Ku-Irtysh, talus, July 17, 1908 — Sap.).

General distribution: East. Sib., Nor. Mong. (Fore Hubs.).

9. **P. pulchrum** Ledeb. Ic. pl. Fl. Ross. impr. Alt. 1 (1829) 20 and Fl. Alt. 4 (1833); 118; DC. Prodr. 6 (1838) 56; Kryl. Fl. Zap. Sib. 11 (1949) 2745; Tzvel. in Fl. SSSR, 21 (1961) 253; Grub. Opred. rast. Mong. [Key to Plants of Mongolia] (1982) 244; Fl. Sin. 76, 1 (1983) 68; Gub. Konsp. fl. Vneshn, Mong. [Conspectus of Flora of Outer Mongolia] (1996) 104. — *Tanacetum pulchrum* (Ledeb.) Sch.-Bip. Tanacet. (1844) 49. — *Chrysanthemum pulchrum* (Ledeb.) Winkl. in Acta Horti Petrop. 10 (1884) 87. — Ic.: Ledeb. Fl. Alt. (1829) l.c. tab. 84; Fl. Kazakhst. l.c. Plate 3, fig. 2.

Described from Altay. Type in St.-Petersburg (LE). Plate II, fig. 4.

On rocky and rubbly slopes of high mountains, alpine grasslands, talus, *Cobresia* meadows, moraines.

IA. Mongolia: *Mong. Alt.* (on slopes of Tsastu-Bogdo mountain range, among clayey-shaly talus, July 31; same site, above Uzun-Dzyur river, among talc-clayey talus, July 31, 1897; mountain slopes toward Nariin river, in upper courses, July 23, 1898 — Klem; basin of Bulugun river, Kharagaitu-khutul' pass, alpine meadows and rubbly plains with snow patches, July 24; Bulugun river basin, upper courses of Ketsu-Sairiin-gol river, along moraines and slopes toward glacier, July 26–1947, Yun; Khasagtu-Khairkhan, nor. slope of Tsagan-Irmyk-ul, in upper courses of Khunker-erin-ama, *Cobresia* grove, Aug. 23, 1972 — Grub., Ulzij. et al; same site, Khasagt-Bogdo-ula mountain, nor. slope, 2900 m, talus, Aug. 8, 1973 — V. Maximowicz; 58 km west of Sagil summer camp, nor. slope of Khargait-ul mountain, 2800 m, in rock crevices, Aug. 8, 1978 — Karam., Beket. et al).

IIA. Junggar ("Nor.-west. part of Sinkiang" — Fl. Sin. l.c.).

General distribution: West. sib. (Altay), Nor, Mong. (Hang.).

10. **P. pyrethroides** (Kar. et Kir.) B. Fedtsch. ex Krasch. in Acta Inst. Bot. Ac. Sci.. URSS, ser. 1, 1 (1933) 176; Kryl. Fl. Zap. Sib. 11 (1949) 2749;

23 Tzvel. in Fl. SSSR, 26 (1961) 256; Ikon. Opred. rast. Pamira [Key to Plants of Pamir] (1963) 234; Fl. Kirgiz. 11 (1965) 125, excl. syn. *P. arassanicum* (Winkl.) O. et B. Fedtsch.; Fl. Kazakhst. 9 (1966) 36; Fl. Sin. 76, 1 (1983) 69; Opred. rast. Sr. Azii [Key to Plants of Mid. Asia] 10 (1993) 608, excl. syn. *P. transiliense* (Herd.) Rgl. et Schmalh. et *P. arassanicum* (Winkl.) O. et B. Fedtsch.— *Richteria pyrethroides* Kar. et Kir. in Bull. Soc. natur. Moscou, 15 (1842) 127; Ledeb. Fl. Ross. 2, 6 (1845) 519. — Ic.: Fl. Kazakhst. l.c. Plate 4, fig. 2; Fl. Sin. l.c. tab. 8, fig. 2.

Described from East. Kazakhstan (Junggar Ala Tau). Type in St.-Petersburg (LE). Map 1.

On rocky slopes and talus in upper belt of mountains, on beds of rivers and brooks.

IB. Kashgar: *West.* (25 km east of Irkeshtam settlement, valley of Sulu-Sakal river, on rocks in rocky gorge, July 26, 1935 — Olsuf'ev).

IIA. Junggar: *Tien Shan* (on road to Kashgar from Urumchi, on exposed slope, 2350 m, No. 6229, July 2; 8 km north of Yakou, on slopes, No. 8490, Aug. 24–1958, A.R. Lee (1959)).

IIIC. Pamir (Kok-Muinak pass, at exit from Tagarma valley, July 27, 1909 — Divn.; Yazlek river, right tributary of Shinba river, Aug.–Sept. 1941 — Serp.).

General distribution: Jung-Tarb., Nor. and Cent. Tien Shan, Mid. Asia (West. Pam.-Alay), West. Sib., Himalayas, China (Nor. and West.).

11. P. richterioides (Winkl.) Krassn. in Opyt ist. razvit. fl. Vost. Tyan'-Shanya [Historical Development of Flora of East. Tien Shan] (1888) 346; Tzvel. in Fl. SSSR, 26 (1961) 255, in nota; Fl. Sin. 76, 1 (1983) 69; excl. syn. *P. karelinii* Krasch. et *Chrysanthemum merzbacheri* B. Fedtsch.— *Chrysanthemum richterioides* Winkl. in Acta Horti Petrop. 10 (1887) 86. — Ic.: Fl. Sin. l.c. tab. 10, fig. 1.

Described from Sinkiang (East. Tien Shan). Type in St.-Petersburg (LE).

On rocky slopes, talus at 1000–3000 m alt.

IIA. Junggar: *Tien Shan* (Dzhagastai, Aug. 11; Talkibash, Aug. 12, 1877; Aryslan, 2700 m, July 13; Kunges, Aug. 22; Kash river, July 3 — 1879, A. Reg., typus!; Tekes, between Kinzu and Kurdai, July 3; Tekes, Koksu river valley, July 8–10 — 1907, Merzb.; Yakou, on slope along margin of talus, No. 4063, Sept. 1, 1957 — Kuan).

General distribution: endemic in Cent. Asia.

12. P. semenovii (Herd.) Winkl. ex O. et B. Fedtsch. in Konsp. Fl. Turk. [Conspectus of Flora of Turkestan] 4 (1911) 186; Tzvel. in Fl. SSSR, 26 (1961) 243; Fl. Kazakhst. 9 (1966) 31; Opred. rast. Sr. Azii [Key to Plants of Mid. Asia] 10 (1993) 606.— *Tanacetum semenovii* Herd. in Bull. Soc. natur. Moscou, 40, 2 (1867) 130.— *Chrysanthemum semenovii* (Herd.) B. Fedtsch. Rast. Turk. [Plants of Turkestan] (1915) 737. — Ic.: Fl. SSSR, l.c. Plate 10, fig. 2; Fl. Kazakhst. 9, Plate 4, fig. 4.

Described from East. Kazakhstan (Trans-Ili Ala Tau). Type in St.-Petersburg (LE).

On rocky slopes and rocks in midbelt of mountains.

IIA. Junggar: *Tien Shan* (pass through Mal. Yuldus, about 2000 m, June 1; Sairam, Talkibash, July 19–1877, A. Reg.; Khanakai south-east of Kul'dzha, June 15, 1878; upper courses of Algoi river, 2000 m, Sept. 12; head of Taldy river, 2200 m, Sept. 15–1879, A. Reg.; Yuldus-Talaty, June 26, 1908 – Merzb.).

General distribution: Fore Balkh., Nor. Tien Shan; endemic in Cent. Asia.

13. P. songoricum Tzvel. in Fl. URSS, 26 (1961) 874, 255; Fl. Kazakhst. 9 (1966) 36; Opred. rast. Sr. Azii [Key to Plants of Mid. Asia] 10 (1993) 607. — Ic.: Fl. Kazakhst. l.c. Plate 8, fig. 2.

Described from East. Kazakhstan (Jung. Ala Tau). Type in St.-Petersburg (LE).

In high-mountain belt on rocky and rubbly slopes, among rocks, on terraces of mountain rivers, above 2000 m alt.

IIA. Junggar: *Tien Shan* (south. Slope of Tien Shan, Sept. 25, 1895 – Rob.; Uch-Turfan, on outcrops in Bedel' pass, July 30, 1908 – Divn.; α 30 km east of Ulugchat, on road to Kensu mine and Kashgar, intermontane plain, July 18, 1959 – Yun. et Yuan'; Turfan, on road from Baiyankhe to Sansanko, No. 5638, June 15; on road to Bogdoshan'-Turfan, on slope above 2000 m, No. 5737, June 18; from Urumchi to Karashar, on terrace, No. 5938, June 21-1958, A.R. Lee (1959); south of Danu village, along course of Danu, in gorge among rocks, No. 366, July 21, 1957 – Kuan; Manas river basin, upper course of Danu-gol river, 1 km south of bend to Se-Daban pass, on rubbly slopes, July 22, 1957, Yun., Lee et Yuan').

IIIC. Pamir (Issyk-su river, 3000-3100 m alt., June 21; same site, Tyngeny-Davan pass, 4200 m, June 26–1942, Serp.).

General distribution: Jung.-Tarb., Nor. Tien Shan; endemic in Cent. Asia.

14. P. transiliense (Herd.) Rgl. et Schmalh. in Acta Horti Petrop. 5, 2 (1878) 618; Tzvel. in Fl. SSSR, 26 (1961) 258; Fl. Kirgiz. 11 (1965) 126; Fl. Kazakhst. 9 (1966) 37; Fl. Sin. 76, 1 (1983) 70. — *P. transiliense* var. *subvillosum* Rgl. et Schmalh. l.c. 618. — *Tanacetum transiliense* Herd. in Bull. Soc. natur. Moscou, 40, 2 (1867) 129. — *Richteria pyrethroides* Kar. et Kir. var. *subvillosa* (Rgl. et Schmalh.) O. et B. Fedtsch. in Konsp. Fl. Turk. [Conspectus of Flora of Turkestan] 4 (1911) 177. — *P. pyrethroides* auct. non (Kar. et Kir.) B. Fedtsch. ex Krasch.: Opred. rast. Sr. Azii [Key to Plants of Mid. Asia] 10 (1993) 608. — Ic.: Fl. Kazakhst. 9, Plate 4, fig. 5; Fl. Sin. l.c. tab. 10, fig. 2.

Described from East. Kazakhstan (Trans-Ili Ala Tau). Type in St.-Petersburg (LE).

On rocky slopes, talus, among rocks in alpine and subalpine mountain belts.

IIA. Junggar: *Tien Shan* (Dzhagastai, June 9, 1882 – A. Reg.; Kshuiku, subalp. belt, Aug. 9, 1913-Knorring).

IIIC. Pamir (Mia gorge, 4000 m, July 21, 1941 – Serp.).

General distribution: Nor. and Cent. Tien Shan, Mid. Asia (West. Tien Shan, Pam.-Alay).

8. Waldheimia Kar. et Kir.
in Bull. Soc. natur. Moscou, 15 (1842) 125. – *Allardia* Decne. in Jacquem. Voy. Ind. Bot. 4 (1844) 87

1. Leaf blade pinnatisected or pinnatifid; ray flowers fertile with distinct pappus; achene pilose .... 2.

+ Leaf blade tri- or palmatilobate at tip; ray flowers sterile, with poorly developed pappus; achene glabrous .... 3.

2. Leaves green, glabrous; anthodium rather small, up to 1 cm in diam. .... 2. W. stoliczkae (Clarke) Ostenf.

+ Leaves greyish green, with loose tomentose pubescence; anthodium large, 1.2–2 cm in diam... .... 3. W. tomentosa (Decne.) Rgl.

3. Entire plant glabrous, green; phyllary cuspidate at tip, entirely glabrous .... 4. W. tridactylites Kar. et Kir.

+ Entire plant greyish or greyish-green, compactly covered with simple long hairs, with loose tomentose pubescence; phyllary somewhat obtuse at tip, with loose tomentose pubescence at base .... l. W. glabra (Decne.) Rgl.

1. W. glabra (Decne.) Rgl. in Acta Horti Petrop. 6, 2 (1880) 309; Tzvel. in Fl. SSSR, 26 (1961) 270; Ikonn. Opred. rast. Pamira [Key to Plants of Pamir] (1963) 237; Fl. Kirgiz. 11 (1965) 128, p.p.; Fl. Sin. 76, 1 (1983) 84; Fl. Xizang. 4 (1985) 728; Opred. rast. Sr. Azii [Key to Plants of Mid. Asia] 10 (1993) 84. – *Allardia glabra* Decne. in Jacquem. Voy. Ind. Bot. 4 (1844) 88. – Ic.: Decne. l.c. tab. 96; Fl. Sin. l.c. tab. 12, fig. 2; Fl. Xizang. l.c. tab. 315, fig. 2.

Described from Himalayas. Type in Paris (P).

On rubbly slopes, fine rocky talus, in rock crevices, 3500–5500 m alt.

IIIB. Tibet: *Chang Tang* ("Getszy", "Shuankhu", "Zhitu" – Fl. Xizang. l.c.).

General distribution: Fore Asia, Mid. Asia (Pam.-Alay), Himalayas, China (South-West.).

2. W. stoliczkae (Clarke) Ostenf. in Hedin, S. Tibet, 6, 3 (1922) 38; Tzvel. in Fl. SSSR, 26 (1961) 268; Fl. Kirgiz. 11 (1965) 129; Fl. Kazakhst. 9 (1966) 39; Fl. Sin. 76, 1 (1983) 83; Fl. Xizang. 4 (1985) 728. – *Allardia stoliczkae* Clarke, Comp. Ind. (1876) 145. – *Waldheimia tomentosa* auct. non (Decne.) Rgl.: Opred. rast. Sr. Azii [Key to Plants of Mid. Asia] 10 (1993) 597. – Ic.: Fl. Kazakhst. l.c. Plate 5, fig. 1.

Described from Himalayas. Type in London (K) or Calcutta (Kolkota) (Cal.). Plate II, fig. 2.

On rocky and rubbly slopes, among rocks, pebble beds of rivers, 3000–5000 m alt.

IIIC. Pamir (Biluli river, at confluence of Chumbus river with it, on bank, among rocks, June 19, 1909 — Divn.).

General distribution: Fore Asia (Afghanistan), Mid. Asia (West. Tien Shan, Pam.-Alay), Himalayas (nor. part), China (Sinkiang).

Note. The author wholly agrees with the view point of N.N. Tzvelev (Fl. SSSR, l.c.) about the independent status of this species. In fact, populations with pubescence as well as with greyish pubescence but not green (characteristic of *W. stoliczkae*) forms of *W. tomentosa* (as pointed out by Tzvelev in Opred. rast. Sr. Azii [Key to Plants of Mid. Asia] l.c.) are found in the same population in the mountains of Mid. Asia. Herbarium material from Himalayas convinced us about the independent status of *W. stoliczkae* because of the stable encounter of green plants without pubescence as well as Pamir *W. stoliczkae* plants which are also glabrous, green or very weakly pubescent.

3. W. tomentosa (Decne.) Rgl. in Acta Horti Petrop. 6, 2 (1891) 372; Tzvel. in Fl. SSSR, 26 (1961) 267; Fl. Kirgiz. 11 (1965) 129; Fl. Kazakhst. 9 (1966) 39; Fl. Sin. 76, 1 (1983) 82; Fl. Xizang. 4 (1985) 728; Opred. rast. Sr. Azii [Key to Plants of Mid. Asia] 10 (1993) 597, excl. syn. *W. stoliczkae* (Clarke) Ostenf. — *Allardia tomentosa* Decne. in Jacquem. Voy. Ind. Bot. 4 (1844) 87. — Ic.: Decne. l.c. tab. 95; Fl. Sin. l.c. tab. 12, fig. 1; Fl. Xizang. 4, tab. 315, fig. 1.

Described from Himalayas. Type in Paris (P). Plate II, fig. 1.

On rocky and rubbly slopes, pebble beds, 3000–4500 m.

IIA. Junggar: *Tien Shan* (Uch-Turfan, Karagaitlik terrain feature above juniper, June 22, 1908 — Divn.).

IIIB. Tibet: *South.* ("Pulan'" — Fl. Sin. l.c.).

General distribution: Fore Asia (Afghanistan), Mid. Asia (Pam.-Alay), Himalayas.

4. W. tridactylites Kar. et Kir. in Bull. Soc. natur. Moscou, 15 (1842) 126; Ledeb. Fl. Ross. 2, 6 (1845) 627; Tzvel. in Fl. SSSR, 26 (1961) 269; Grub. Opred. rast. Mong. [Key to Plants of Mongolia] (1982) 244; Opred. rast. Sr. Azii [Key to Plants of Mid. Asia] 10 (1993) 598; Gub. Konsp. fl. Vneshn. Mong. [Conspectus of Flora of Outer Mongolia] (1996) 109. — *Allardia tridactylites* (Kar. et Kir.) Sch.-Bip. in Pollichia, 20, 21 (1863) 442; Kryl. Fl. Zap. Sib. 11 (1949) 2759. — Ic.: Fl. Kazakhst. l.c. Plate 5, fig. 2.

Described from East. Kazakhstan (Junggar Ala Tau). Type in St.-Petersburg (LE). Map 2.

On rocky, rocky-stony slopes of high mountains, rubbly talus, rocky placers, along floor of gorges, 3000–4000 m alt.

IA. Mongolia: *Khobd.* (30 km from Achit-nur lake, Kharkhira foothills, south-west. slope toward left bank of Irtin-gol river, among rocks, 3200 m, June 16, 1978 — Karam., Beket. et al), *Mong. Alt.* (waterdivide between Tsagan-gol and Aksu rivers, rubbly alpine tundra, July 21, 1909—Sap.; Tolbo-Kungei mountain range (central portion), alpine belt, rocky, semi-turf-covered placers, Aug. 5, 1945; Adzhi-Bogdo mountain range, Burgasyndava pass, between Indertiin-gol and Dzuslangin-gol, rubbly placers, Aug. 2, 1947—Yun.; Adzhi-Bogdo mountain range, south, slope toward Ikhe-gol river, rocky-stony slope, Aug. 6, 1979—Grub., Dar., Muld.), *Gobi Alt.* (Baga-Bogdo mountain, at border of vegetation, July 30, 1895—Klem.; Ikhe-Bogdo mountain range, south. slope, Bityutenama creek valley, alpine zone, Aug. 12, 1927—Simukova; Ikhe-Bogdo mountain range, south. slope, upper Baga-Artsatuin-ama creek valley, among rocky placers, 3900 m, Sept. 8, 1943; Ikhe-Bogdo mountain range, rubbly placers, June 29, 1945—Yun.; Ikhe-Bogdo mountain range, south. slope, on talus among rocks, 3500 m, Aug. 4, 1973—Isach. et Rachk.).

IIA. Junggar: *Tien Shan* (Kumbel', 3000 m, June 31; Monguto, 3000 m, June 4; Kunges, Aryslyn-Daban area, Aug. 22–1879, A. Reg.; Uch-Turfan, Kara-Garlin gorge, above juniper, June 22, 1908—Divn.; Manas river basin, upper courses of Danu-gol river near site of ascent to Se-Daban pass, along floor of gorge, July 22, 1957—Yun., Lee et Yuan').

General distribution: Jung.-Tarb., Nor. and Cent. Tien Shan, Mid. Asia (West. Tien Shan, Pam.-Alay), Nor. Mong. (Hang., Fore-Hubs.).

9. Cancrinia Kar. et Kir.
in Bull. Soc. natur. Moscou, 15 (1842) 124

1. Annual-biennial plants with roots easily pulled out; thalamus highly convex; conical-hemispherical; achene glabrous .... 2. C. discoidea (Ledeb.) Poljak. ex Tzvel.

+ Perennial plants, often forming compact or loose mats; thalamus moderately covex, achenes weakly pubescent .... 2.

2. Plants 30–50 cm tall with many lignifying shoots; lower cauline leaves on petioles as long as their blades, greyish green .... 4. C. maximowiczii Winkl.

+ Perennial plant, often forming compact or loose mats; thalamus moderately convex, achenes faintly pubescent .... 3.

3. Anthodium 7–12 mm in diam.; phyllary narrow-lanceolate, cuspidate at tip; thalamus moderately covex, achenes with highly elongated crown, pilose all over surface .... 3. C. lasiocarpa Winkl.

+ Anthodium 10–17 mm in diam.; phyllary oblong, obtuse at tip; thalamus subconvex, achene diffusely pilose with somewhat elongated crown .... 4.

4. Entire phyllary with broad dark brown margin, enlarged at tip of leaflet, achene diffusely pilose only in upper part .... l. C. chrysocephala Kar. et Kir.

+ Entire phyllary with narrow light brown margin, not enlarged at tip of leaflet, achene diffusely pilose all over surface .... 5. C. tianschanica (Krasch.) Tzvel.

1. C. chrysocephala Kar. et Kir. in Bull. Soc. natur. Moscou, 15 (1842) 125; Ledeb. Fl. Ross. 2, 6 (1845) 519; Krasch. in Not. syst. Herb. Horti Bot. Petrop. 3 (1922) 81, excl. ssp. *tianschanica*; Poljak. in Not. syst. (Leningrad), 19 (1959) 370, excl. syn. ssp. *tianschanica*; Tzvel. in Fl. SSSR, 26 (1961) 314; Fl. Kazakhst. 9 (1966) 50; Fl. Sin. 76, 1 (1983) 100; Opred. rast. Sr. Azii [Key to Plants of Mid. Asia] 10 (1993) 534. — *Allardia chrysocephala* (Kar. et Kir.) Sch.-Bip. in Pollichia, 20, 21 (1863) 442. — Ic.: Fl. SSSR, l.c. Plate 13, fig. 1; Fl. Sin. l.c. tab. 15, fig. 2.

Described from East. Kazakhstan (Junggar Ala Tau), Type in St.-Petersburg (LE).

On rubbly slopes, fine-rubbly talus, pebble beds of river valleys, 3000–4000 m alt.

IIA. Junggar: *Tien Shan* (basin of Manas river, upper courses of Ulan-usu river before ascent to Danu pass, on pebble beds along valley floor, July 19; same site, valley of Danu-gol river at Se-Daban pass, in valley of upper Manas, fine-rubbly talus near glacier, July 21–1957, Yun., Lee et Yuan'; 30 km from Nyutsyuan'tsza, along bank of Ulan-usu river, No. 318, July 19, 1957 — Kuan).

General distribution: Jung.-Tarb., Nor. Tien Shan; endemic in Cent. Asia.

2. C. discoidea (Ledeb.) Poljak. ex Tzvel. in Fl. SSSR, 26 (1961) 313; Grub. Opred. rast. Mong. [Key to Plants of Mongolia] (1982) 244; Fl. Sin. 76, 1 (1983) 99; Fl. Intramong. ed. 2, 4 (1993) 583; Gub. Konsp. fl. Vneshn. Mong. [Conspectus of Flora of Outer Mongolia] (1996) 100. — *Pyrethrum discoideum* Ledeb. Fl. Alt. 4 (1833) 119; DC. Prodr. 6 (1838) 59; Ledeb. Fl. Ross. 2, 6 (1845) 556. — *Matricaria ledebourii* (Sch.-Bip.) Schischk. in Kryl. Fl. Zap. Sib. 11 (1949) 2743; Grub. in Not. syst. (Leningrad), 19 (1959) 5, 51. — *Chrysanthemum ledebourii* (Sch.-Bip.) Ling in Contr. Inst. Bot. Nat. Ac. Peiping, 3 (1935) 474. — Ic.: Ledeb. Icon. pl. Fl. Ross. impr. Alt. (1830), tab. 153; Fl. Sin. l.c. tab. 15, fig. 1; Fl. Intramong. l.c. tab. 231, fig. 1.

Described from Altay. Type in St.-Petersburg (LE). Map 2.

On rocky, clayey-rubbly slopes of low mountains, in saltwort barren lands, along floor of gorges.

IA. Mongolia: Common in *Mong. Alt.* (east), *Val. Lakes, Gobi-Alt., East Gobi* (Ulan-Osh), *West. Gobi, Alash. Gobi, Ordos, Khesi.*

IIA. Junggar: *Jung. Gobi* (nor.-west. Junggar plain near Baikoutszy settlement on road between Karamai and Orkha, among rocky deserts, June 25, 1957 — Yun., Lee et Yuan'; 33 km east of Shikho town, wormwood-saltwort desert nor. of Tien Shan foothills, Aug. 30, 1959 — Petr.; 7 km south-east of Ushig border post, vicinity of Ushigiin-us spring and Budun-ul mountains, July 29, 1988 — Gub., Dar. et Kam.).

General distribution: Aralo-Casp., Fore Balkh., Jung.-Tarb.; West. Sib. (Altay).

3. C. lasiocarpa Winkl. in Acta Horti Petrop. 12 (1892) 30; Krasch. in Not. syst. (Leningrad), 3 (1922) 84; Fl. Sin. 76, 1 (1983) 100.

Described from Qinghai (Nanshan). Type in St.-Petersburg (LE).

On arid mountain slopes.

IA. Mongolia: *Alash. Gobi* ("Nin'sya" – Fl. Sin. l.c.).

IIIA. Qinghai: *Nanshan* (Nanshan foothills, June 22, 1879, No. 248 – Przew., typus!; nor. Nanshan foothills, clay with pebbles, May 11, 1894–Rob.).

General distribution: endemic in Cent. Asia.

28 4. **C. maximowiczii** Winkl. in Acta Horti Petrop. 12 (1892) 29; Krasch. in Not. syst. Herb. Horti Bot. Petrop. 3 (1922) 81; Poljak. in Not. syst. (Leningrad) 19 (1959) 372; Fl. Sin. 76, 1 (1983) 98; Drevesn. rast. Tsinkhaya [Woody Plants of Qinghai] (1987) 626; Fl. Intramong. ed. 2, 4 (1993) 583. – Ic.: Drevesn. rast. Tsinkhaya [Woody Plants of Qinghai] l.c. tab. 446.

Described from Qaidam. Type in St.-Petersburg (LE).

On clayey and sandy-clayey slopes of low mountains.

IA. Mongolia: *Alash. Gobi* ("Alashan" mountain range – Fl. Sin. l.c.; Fl. Intramong. l.c.).

IC. Qaidam: *Mount.* (Qaidam, Aug. 5, 1879, No. 397 – Przew., typus!).

General distribution: endemic in Cent. Asia.

5. **C. tianschanica** (Krasch.) Tzvel. in Fl. URSS, 26 (1961) 315; Fl. Kirgiz. 11 (1965) 140; Fl. Sin. 76, 1 (1983) 102; Opred. rast. Sr. Azii [Key to Plants of Mid. Asia] 10 (1993) 535. – *C. chrysocephala* Kar. et Kir. ssp. *tianschanica* Krasch. in Not. syst. Herb. Horti Petrop. 3 (1922) 81. – *C. chrysocephala* auct. non Kar. et Kir.; Poljak. in Not. syst. (Leningrad), 19 (1959) 370, quoad syn. *C. chrysocephala* ssp. *tianschanica* et pl. e Tian-Schan.

Described from Kirghizia (Cent. Tian shan). Type in St.-Petersburg (LE). Plate II. Fig. 5.

On rubbly and clayey-rubbly slopes of high mountains, 3000 m and above.

IIA. Junggar: *Jung. Gobi* (7 km south-east of Urumchi on road to Turfan, second terrace on right of Urumchinka river, nanophyte desert, July 31, 1957 – Yun., Lee et Yuan').

General distribution: Jung.-Tarb., Nor. and Cent. Tien Shan; endemic in Cent. Asia.

10. Tanacetum L.

Sp. pl. (1753) 843, p.p. and Gen. pl. ed. 5 (1754) 366

1. Plants 30–150 cm tall; fertile shoots profusely leafy, without contracted vegetative shoots; achenes 1.2–2.5 mm long .... 2.
+ Plants up to 40 (60) cm tall; fertile shoots sparsely leafy, many vegetative shoots in the form of rosettes of radical leaves; achenes 1.5–3.5 mm long .... 3.

2. Anthodium 5–8 mm in diam.; phyllary with narrow light-coloured or brownish margin; achenes 1.2–1.8 mm long ....
.... 5. T. vulgare L.

+ Anthodium 7–13 mm in diam.; phyllary with relatively broad dark brown margin; achenes 1.5–2 mm long ....
.... 1. T. boreale Fisch. ex DC.

3. Anthodium 4–7 mm in diam.; marginal pistillate flowers varying from tubular to ligulate, their limb not more than 1.5 mm in diam.; outer phyllary broad-ovoid to lanceolate-ovoid ....
.... 3. T. santolina Winkl.

+ Anthodium invariably with ray flowers considerably longer than disc flowers; limb of corolla of ray flowers longer than 1.5 mm; outer phyllary lanceolate, subacute .... 4.

4. Plants usually greyish green due to fairly profuse pubescence; anthodium 5 (6)–8 (10) mm in diam., on relatively slender pedicels, aggregated into fairly compact corymb ....
.... 4. T. tanacetoides (DC.) Tzvel.

+ Plants usually dull green due to rather sparse pubescence; anthodium (6) 8–10 (12) mm in diam., aggregated on fairly thickened pedicels into very compact, subcapitate corymb ....
.... 2. T. crassipes (Stschegl.) Tzvel.

1. **T. boreale** Fisch. ex DC. Prodr. 6 (1838) 128; Tzvel. in Fl. SSSR, 26 (1961) 327; Fl. Kirgiz. 11 (1965) 143; Fl. Kazakhst. 9 (1966) 57; Opred. rast. Sr. Azii [Key to Plants of Mid. Asia] 10 (1993) 611. — *T. vulgare* L. var. *boreale* (Fisch. ex DC.) Trautv. et Mey. Fl. Ochot. (1856) 54; Trautv. in Bull. Soc. natur. Moscou, 39, 1 (1866) 359; Kryl. Fl. Zap. Sib. 11 (1949) 2757. — Ic.: Gmel. Fl. Sib. 2 (1749) tab. 65, fig. 1; Fl. Kazakhst. 9, Plate 7, fig. 1.

Described from garden specimens, probably from Far East (Fl. SSSR, l.c.). Type in Berlin (B).

In larch and aspen forests, forbs meadows, among shrubs.

IA. Mongolia: *Mong. Alt.* (nor. slope of Khara-Adzarga mountain range, around Khairkhan-Duru river, larch forest, Aug. 25, 1930–Pob.), *Cent. Khalkha* (bank of Uber-Dzhargalante river, near Botoga mountain, aspen forest, Oct. 21, 1925 — Krasch. et Zam.).

General distribution: Jung.-Tarb., Cent. Tien Shan, West. and East. Sib., Far East, Nor. Mong. (Hent., Hang.), China (Nor.), Korean peninsula, Nor. Amer. (Alaska).

2. **T. crassipes** (Stschegl.) Tzvel. in Fl. SSSR, 26 (1961) 338; Fl. Kazakhst. 9 (1966) 58; Fl. Sin. 76, 1 (1983) 78; Opred. rast. Sr. Azii [Key to Plants of Mid. Asia] 10 (1993) 612. — *Pyrethrum crassipes* Stschegl. in Bull. Soc. natur. Moscou, 27 (1854) 172. — Ic.: Fl. Kazakhst. l.c. Plate 7, fig. 3; Fl. Sin. l.c. tab. 11, fig. 1.

Described from Altay (Narym mountain range). Type in Moscow (MW).

On rocky arid steppe mountain slopes, rarely on hummocky sand.

IIA. Junggar: *Zaisan* (left bank of Chern. Irtysh river, Mai-Kain river, hummocky sand, July 7, 1914–Schischk.).

General distribution: Jung.-Tarb.; West. Sib. (Altay).

Note. This species is closely related to *T. tanacetoides* (DC.) Tzvel. but differs from it in anthodia on thick pedicels with tomentose pubescence, gathered into compact subcapitate corymb.

3. **T. santolina** Winkl. in Acta Horti Petrop. 11, 12 (1891) 375; Tzvel. in Fl. SSSR, 26 (1961) 343; Fl. Kazakhst. 9 (1966) 61; Fl. Sin. 76, 1 (1983) 78; Opred. rast. Sr. Azii [Key to Plants of Mid. Asia] 10 (1993) 613. — *Pyrethrum kasachstanicum* Krasch. in Not. syst. (Leningrad), 9 (1946) 160; Kryl. Fl. Zap. Sib. 11 (1949) 2753. — Ic.: Krasch. l.c. Plate 3.

Described from West. Kazakhstan (Kzyl-Orda province). Type in St.-Petersburg (LE).

On clayey, sandy, solonetzic soils and solonetzes in steppe and desert zones.

IIA. Junggar: *Jung. Gobi* ("Nor. Sinkiang" — Fl. Sin. l.c.).

General distribution: Aralo-Casp., Fore-Balkh., West. Sib. (south-west.).

Note. One more species *T. scopulorum* (Krasch.) Tzvel. has been cited (Fl. Sin. l.c.) for Sinkiang although it hardly grows in that territory. Fl. SSSR and Opred. rast. Sr. Azii [Key to Plants of Mid. Asia] regard *T. scopulorum* as endemic in Balkhash region and Betpakdala.

4. **T. tanacetoides** (DC.) Tzvel. in Fl. SSSR, 26 (1961) 337; Fl. Kazakhst. 9 (1966) 58; Grub. Opred. rast. Mong. [Key to Plants of Mongolia] (1982) 244; Fl. Sin. 76, 1 (1983) 77; Opred. rast. Sr. Azii [Key to Plants of Mid. Asia] 10 (1993) 612; Gub. Konsp. fl. Vneshn. Mong. [Conspectus of Flora of Outer Mongolia] (1996) 107. — *Pyrethrum tanacetoides* DC. Prodr. 6 (1838) 59; Ledeb. Fl. Ross. 2, 6 (1845) 555; Kryl. Fl. Zap. Sib. 11 (1949) 2753. — *P. millefolium* Willd. var. *tanacetoides* (DC.) Trautv. in Bull. Soc. natur. Moscou, 39, 1 (1866) 348. — *Tanacetum tauricum* Sch.-Bip. var. *tanacetoides* Trautv. ex Herd. in Bull. Soc. natur. Moscou, 40, 2 (1876) 128. — *Pyrethrum millefolium* auct. non Willd.: Ledeb. Icon. pl. Fl. Ross. 4 (1833) 20, tab. 369.

Described from south. Altay. Type in Geneva (G). Isotype in St.-Petersburg (LE).

On arid rocky slopes in mountain steppes, up to 2000 m alt.

IA. Mongolia: *Mong. Alt.* (upper course of Bain-gol river, right bank tributary of Bulugun river, mountain steppe, July 23, same site, Bain-gol river, on road to summer camp at Bulgan somon, steppe slopes, July 23–1947, Yun.; Sangiin-gol river basin, upper course of Chern. Irtysh, near Chinese border, 25 km south-east of Dayan-nur post, 1800 m, July 16–17, 1988-Kam., Gub., Dar.).

IIA. Junggar: "*Sinkiang*"-Fl. Sin. l.c.

General distribution: Jung.-Tarb., West. Sib., Nor. Mong. (Hent., Hang.).

5. T. vulgare L. Sp. pl. (1753) 845; Kryl. Fl. Zap. Sib. 11 (1949) 2756; Tzvel. in Fl. SSSR, 26 (1961) 326; Fl. Kirgiz. 11 (1965) 143; Fl. Kazakhst. 9 (1966) 56; Grub. Opred. rast. Mong. [Key to Plants of Mongolia] (1982) 244; Fl. Sin. 76, 1 (1983) 76; Fl. Intramong. ed. 2, 4 (1993) 577; Opred. rast. Sr. Azii [Key to Plants of Mid. Asia] 10 (1993) 611; Gub. Konsp. fl. Vneshn. Mong. [Conspectus of Flora of Outer Mongolia] (1996) 107. — *T. vulgare* var. *genuinum* Trautv. in Bull. Soc. natur. Moscou, 39, 1 (1866) 359. — *Chrysanthemum vulgare* (L.) Bernh. Verz. Pflanz. Erfurt (1800) 144. — *Ch. vulgare* var. *boreale* (Fisch. ex DC.) Makino et Nemoto, Fl. Japon. (1925) 43; Kitam. in Mem. Coll. Sci. Kyoto Univ. ser. B, 15 (1940) 342.

Described from West. Europe. Type in London (Linn.).

In larch forests and their borders, meadows, glades, in steppes, scrubs, along banks of rivers, as weed on fallow lands, along borders of fields, on roadsides. Widely distributed species.

IA. Mongolia: *Khobd.* (Kharkira mountain group, pass to Talsa spring, July 30, 1930- Gr.-Grzh.; 20 km nor.-west of Sagil, Ul'd-ziityn area, scrubs, June 29, 1978 — Karam., Beket. et al), *Mong. Alt.* (midbelt of mountains near Khalyun, steppe, Aug. 24, 1943; nor. trail of Taishiri-ul, along floor of sand ravine, July 11, 1945 — Yun.), *East. Mong.* (Khalkhin-gol river valley, 13 km south-east of Khamar-Daban, willow groves, Aug. 11, 1949 — Yun.).

IIA. Junggar: *Dzhark.* (Kul'dzha, July 13, 1875 — Lar.; Pilyuchi river near Kul'dzha, July 23, 1878 — A. Reg.; Pilyuchi river, June 21, 1879 — Fet.), *Tien Shan* (Sairam-nur, July 12, 1877; Kash river, Aug. 18, 1878; Nilki gorge, about 2000 m, June 26; Nilki mountain range, June 26–1879, A. Reg.; Mukhurdai river, June 19, 1893 — Rob.; south-east. part of Ketmen' mountain range, Sarbushin pass on road to Kzyl-Kure from Ili, steppified meadow on nor. slope, Aug. 7, 1957 — Yun., Lee, Yuan').

General distribution: Fore-Balkh., Aralo-Casp., Jung.-Tarb., Nor. and Cent. Tien Shan, Europe, West. and East. Sib., Far East, Nor. Mong. (all regions), Korean peninsula, Japan, Nor. Amer.

11. Poljakovia Grub. et Filat.

Novit. syst. pl. vasc. 33 (2001) 227.

— *Tanacetum* auct. non L.

1. Fertile shoots rod-like, compactly leafy; anthodium broad-ovoid, phyllary with broad dark brown margin ....
    .... 2. P. falcatolobata (Krasch.) Grub. et Filat.

+   Fertile shoots thick, coarse, rather sparsely leafy; anthodium ovoid, with narrow brownish margin .... 2.

2. Lower cauline leaves and leaves of barren shoots pinnatifid, their lateral segments entire, oblong-lanceolate, subobtuse at tip .... 3. P. kaschgarica (Krasch.) Grub. et Filat.

+ Lower cauline leaves and leaves of barren shoots pinnatisected, their lateral segments linear, short-cuspidate at tip ....
.... 1. P. alashanensis (Ling) Grub. et Filat.

1. P. alashanensis (Ling) Grub. et Filat. in Novit. syst. pl. vasc. 33 (2001) 227.— *Tanacetum alashanense* Ling in Contr. Inst. Bot. Nat. Ac. Peiping, 2 (1935) 502.— *Hippolytia alashanensis* (Ling) Ling in Acta Phytotax. Sin. 17, 4 (1979) 63; Fl. Sin. 76, 1 (1983) 89; Fl. Intramong. ed. 2, 4 (1993) 578. —Ic.: Fl. Intramong. l.c. tab. 228, fig. 4–6.

Described from Nins'ya province (Alashan mountain range). Type in Beijing (PE).

On clayey and rubbly slopes.

IA. Mongolia: *Alash. Gobi* ("Nin'sya, Khelanshan'", Fl. Sin. l.c.).

General distribution: endemic.

Note. The name of this species, as seen from synonymy, "migrated" from one genus to another. On the strength of similar morphological features that are also characteristic of *P. kaschgarica* (Krasch.) Grub. et Filat., we have placed it in genus *Poljakovia* Grub. et Filat.

2. P. falcatolobata (Krasch.) Grub. et Filat. in Novit. syst. pl. vasc. 33 (2001) 227.— *Tanacetum falcatolobatum* Krasch. in Not. syst. (Leningrad), 4, 1 (1923) 7.

Described from Qinghai (Nanshan). Type in St.-Petersburg (LE). Plate III, fig. 2. Map 2.

On arid clayey slopes, in rock crevices, 1800–2600 m alt.

IIIA. Qinghai: *Nanshan* (path below along Rangkhta-gol river up to Tetung river bank, July 29, 1872—Przew., typus!; South-Kukunor mountain range, along arid slopes, 1800–2400 m, June 26, 1894—Rob.); *Amdo* (upper course of Huang He, Kha-Gomn, on clayey slopes, May 28, 1880—Przew.).

IIIB. Tibet: *Weitzan* (Burkhan-Budda mountain range, Khatu gorge, 2600 m, arid slopes, in rock crevices, July 25, 1901—Lad.).

General distribution: endemic in Cent. Asia.

3. P. kaschgarica (Krasch.) Grub. et Filat. in Novit. syst. pl. vasc. 33 (2001) 227.— *Tanacetum kaschgaricum* Krasch. in Acta Inst. bot. Ac. Sci. URSS, 1, 1 (1933) 175.— *Hippolytia kaschgarica* (Krasch.) Poljak. in Not. syst. (Leningrad), 18 (1957) 220.

Described from Sinkiang. Type in St. –Petersburg (LE). Plate III, fig. 1. Map 2.

On arid rocky slopes of mountains.

IB. Kashgar: *Nor.* (Ishma mountains, between Kucha and Karashar, Aug. 22, 1929— M. Pop., typus !).

General distribution: endemic in Cent. Asia.

Note. This species was described for the first time by I.M. Krascheninnikov as *Tanacetum kaschgaricum* Krasch. (in Acta ... l.c.). Later, based on the homogamous state

of anthodium, P.P. Poljakov placed it in his new genus *Hippolytia* Poljak. (in Not. ... l.c.)
N.N. Tzvelev excluded this species from the above genus in his treatment for "Flora
SSSR" [Flora of USSR] as extremely isolated with respect to other species (intense
lignification, shape and division of leaf blade, size and shape of anthodia and phyllary).
Based on these characteristics, we too included this species in the new genus *Poljakovia*
Grub. et Filat.

12. Chrysanthemum L.

Sp. pl. (1753) 887, p.p.—*Dendranthema* (DC.) Des Moul. in Actes Soc.
Linn. Bordeaux, 20 (1855) 561, p.p.; Tzvel. in Fl. URSS. 26 (1961) 364,
p.p.—*Pyrethrum* sect. *Dendranthema* DC. Prodr. 6 (1838) 62.— *Chrysanthemum* sect. *Pyrethrum* subsect. *Dendranthema* (DC.) Kitam. in Acta
Phytotax. Geobot. 4 (1935) 36 and Compos. Japon. 2 (1940) 350, p.p.

1. Small subshrubs with thick multicipital woody root; fertile
   shoots lignifying at base; lower cauline leaves whitish due to
   thin adherent fine tomentum, subpilose above with perceptible
   punctate glands .... 3. C. sinuatum Ledeb.

+ Perennial plant with fine rhizome; fertile shoots herbaceous;
   lower cauline leaves green, subglabrous, with many punctate
   glands on both sides .... 2.

2. Radical and lower cauline leaves pinnati- or palmatilobed for
   not more than up to middle (rarely slightly deeper), with broad
   ovoid or broad-oblong lobes mostly perceptibly dentate along
   margin .... 2. C. naktongense Nakai.

+ Radical and lower cauline leaves simply or palmately partite up
   to middle (generally up to narrow-winged axis) or divided into
   narrow-linear or linear-oblong lobes which in turn are divided
   into much shorter lobules .... 3.

3. Plant 10–25 cm tall; lower cauline leaves on stalks not enlarged
   toward base (without wings); terminal leaf lobules narrow-
   linear, up to 1.5 mm broad; anthodium 2–3 cm in diam. ....
   .... 1. C. chalchingolicum Grub.

+ Plant 50 cm tall; lower cauline leaves on winged stalks; terminal
   leaf lobules oblong, 2–3 mm broad; anthodium much larger, 3–6
   cm in diam. .... 4. C. zawadskii Herbich.

1. C. chalchingolicum Grub. in Bot. zhurn. 57, 12 (1972) 1592; id.
Opred. rast. Mong. [Key to Plants of Mongolia] (1982) 243.—
Dendranthema chalchingolicum (Grub.) Gubanov et R. Kam. in Bull. Soc.
natur. Moscou, 101, 2 (1996) 62; Gub. Konsp. fl. Vneshn. Mong.
[Conspectus of Flora of Outer Mongolia] (1996) 101. — Ic.: Grub. Opred.
rast. Mong. [Key to Plants of Mongolia] l.c. Plate 80, fig. 592.

Described from Mongolia (Khalkhin-gol river). Type in St.-Petersburg (LE).

On steppe slopes of low mountains, solonetzic arid common cattail steppes.

IA. Mongolia: *East. Mong.* (Khalkhin-gol river, Sumber terrain feature, steppe, Sept. 4, 1928 — Tug.; Khalkhin-gol river 12 km south of Khukh-Undur-obo, slope on left bank (opposite common grave of 1939 War), arid solonetzic common cattail — cereal grass steppe, Aug. 15, 1970, No. 694 — Grub., Ulzij., Tserenbalzhid, typus!; Khalkhin-gol river, 46 km south-east of Khamar-Daba settlement, sedge — cereal grass — forbs steppe, July 27, 1970 — Karam., Safronova).

General distribution: endemic.

2. C. naktongense Nakai in Bot. Mag. Tokyo, 23 (1909) 126. — *Leucanthemum sibiricum* DC. var. *latilobum* Maxim. Primit. Fl. Amur. (1859) 156. — *Chrysanthemum sibiricum* (DC.) Kom. var. *latilobum* (Maxim.) Kom. in Acta Horti Petrop. 25 (1907) 642; Kom. et Alis. Opred. rast. Dal'nevost. kraya [Key to Plants of Far Eastern Region] 2 (1932) 1031. — *Ch. zawadskii* Herbich subsp. *latilobum* (Maxim.) Kitag. Lin. fl. Manch. (1939) 444. — *Dendrathema naktongense* (Nakai) Tzvel. in Fl. URSS, 26 (1961) 375; Fl. Sin. 76, 1 (1983) 34; Fl. Intramong. ed. 2, 4 (1993) 574. — Ic.: Kom. et Alis. Opred. rast. Dal'novost. kraya [Key to Plants of Far Eastern Region] Plate 307.

Described from Korea. Type in Tokyo (TI).

On rubbly slopes of mountains, in steppes, willow groves, rarely on sand.

IA. Mongolia: *East. Mong.* (vicinity of Manchuria station, on rocks, Aug. 22, 1902 — Litw.; same site, 1915 — Nechaeva; Sukhe-Bator, Dzodol-khan somon, Unetin-Suburga-obo area, rubbly slopes, Sept. 15, 1949 — Yun.; Dariganga, Shilin-Bogdo-ula, along slopes of extinct volcano, Aug. 11, 1970 — Grub., Ulzij., Tserenbalzhid; "Shilingol" ajmaq [administrative territorial unit in Mongolia], "Khukh-khoto town", "Ulantsab" — Fl. Intramong. l.c.).

General distribution: Far East, China (Dunbei), Korean peninsula.

3. C. sinuatum Ledeb. Fl. Alt. 4 (1933) 116; Grub. Opred. rast. Mong. [Key to Plants of Mongolia] (1982) 243. — *Leucanthemum sinuatum* (Ledeb.) DC. Prodr. 6 (1838) 46; Ledeb. Fl. Ross. 2, 6 (1845) 542; Kryl. Fl. Zap. Sib. 11 (1949) 2742. — *Dendranthema sinuatum* (Ledeb.) Tzvel. in Fl. URSS, 26 (1961) 370; Gub. Konsp. fl. Vneshn. Mong. [Conspectus of Flora of Outer Mongolia] (1996) 101. — Ic.: Ledeb. Ic. Pl. Fl. Ross. 5 (1834) tab. 494.

Described from Altay. Type in St.-Petersburg (LE). Plate I, fig. 4.

On rocky slopes, rocks, in middle and upper belts of mountains.

IA. Mongolia: *Khobd.* (Gub. l.c.), *Mong. Alt.* (basin of Dzhelta river, valley of Korumdy-gol river, near Chinese border, Aug. 9, 1979 — Grub.; valley of Ëlt river, 30 km south of Altay somon, rocky slopes, 1700–2000 m, July 15; basin of Ëlt river (upper courses of Chern. Irtysh), near Chinese border, July 22 — 1988, Kam., Gub., Dar.).

IIA. Junggar: *Cis-Alt.* (vicinity of Fuyun', Aug. 17, 1956 – Ching).
General distribution: endemic in Altay.

4. C. zawadskii Herbich, Addit. Fl. Galic. (1831) 44; DC. Prodr. 6 (1838) 67; S.J. Hu in Quart. Journ. Taiwan Mus. 19 (1966) 45; Grub. Opred. rast. Mong. [Key to Plants of Mongolia] (1982) 243. – *Chrysanthemum sibiricum* Fisch. ex Turcz. Fl. Baic.-Dahur. 2 (1856) 42, in syn. – *Ch. sibiricum* var. *acutilobum* (DC.) Kom. et Alis. Opred. rast. Dal'nevost. kraya [Key to Plants of Far East. Region] 2 (1932) 1031. – *Ch. hwang-shanense* Ling in Contr. Inst. Bot. Ac. Peiping, 3 (1935) 472. – *Leucanthemum sibiricum* auct. non DC.: Ledeb. Fl. Ross. 2, 6 (1845) 541; Kryl. Fl. Zap. Sib. 11 (1949) 2741. – *Chrysanthemum naktongense* Nakai var. *dissectum* (Ling) Hand.-Mazz. in Acta Horti Gothob. 12 (1938) 256. – *Dendranthema zawadskii* (Herb.) Tzvel. in Fl. URSS, 26 (1961) 376; Fl. Sin. 76, 1 (1983) 45; Fl. Intramong. ed. 2, 4 (1993) 571; Gub. Konsp. fl. Vneshn. Mong. [Conspectus of Flora of Outer Mongolia] (1996) 101. – Ic.: Fl. Intramong. l.c. tab. 224, fig. 5–9.

Described from Europe (Carpathians). Type in Wien (Vienna) (W). Plate I, fig. 3.

On rubbly, limestone and sandy soils in steppe and forest-steppe zones.

IA. Mongolia: *Cent. Khalkha* (Munkh-Khan-ula mountain, 1607 m, cereal grass mountain steppe, July 25, 1974 – Golubk. et Tsogt), *East. Mong.* (between Kulusutai and Dolon-Nor, 1870, No. 83 – Lom.; Shara-Muren, 1925, No. 481-Chaney; Boro-Khurkha, Aug. 25, 1927 – Terekhovko; 25 km south of Bain-Dung somon, near Irgai-ula pass, cereal grass – tansy steppe, Aug. 26; Bain-Dung somon, 7 km north of Chingis-Khan embankment on road to Choibalsan, forbs – cereal grass steppe, Aug. 26-1949, Yun.; Ara-Zhargalant somon, Aug. 25, 1954; Kharkhira river, south. Bayan-ul somon, south. slope of mountain, blue grass – tansy steppe, Aug. 2; Khotono, 5 km west of Morgain-bulak, south-west. slope of hillock, sheep's fescue-astro-wormwood steppe, Aug. 17, 1956 – Dashnyam; vicinity of Khailar town, alliaceous steppe, 1959 – L. Ivanov; Sukhe-

34 Bator ajmaq [administrative territorial unit in Mongolia], 33 km north-west of Khutliin-khuduk, hummocky area, Aug. 14, 1971 – Dashnyam, Karam. et Safronova; between Gurban-Dzagal somon and Dornot settlement, Aug. 2, 1985 – Kam.; westward of road between Batnorov and Norovlin somons, 35 km from latter, sparse birch forest with willow in upper portion of nor. slope, Aug. 11, 1989 – Grub., Gub., Dar.; "Shilin-Khoto" – Fl. Intramong. l.c.).

General distribution: Arctic, Europe, East. Sib., Far East, Nor. Mong. (Fore Hubs., Hent., Hang.), China (Dunbei).

Note. Highly polymorphous species. Over its fairly extensive distribution range, it varies in flower characteristics (mostly in the colour of ray flowers) and leaves, specially with respect to the pubescence of leaf blade and size and breadth of terminal leaf lobes.

The reference cited in Fl. Sin. 76, 1 (1983) and Fl. Intramong. 2, 4 (1993) to the find of *C. maximowiczii* Kom. in East. Mongolia (Ulantsab, Khukh-Khoto, Baotou) has not been confirmed by our herbarium material.

13. Brachanthemum DC.

DC. Prodr. 6 (1838) 44. — *Chrysanthemum* sect. *Argyranthemum* (Web. ex Sch.-Bip.) Benth. ex Hook. f. Gen. Pl. 2 (1876) 426, p.p.; O. Hoffm. in Engl. et Prantl. Pflanzen-fam. 4, 5 (1889) 278. — *Chrysanthemum* sect. *Argyranthemum* subsect. *Brachanthemum* (DC.) Ling in Contr. Inst. Bot. Nat. Ac. Peiping, 3 (1935) 476.

1. Small subshrubs up to 15 cm tall, with highly branched lignifying thin shoots; leaf blade 2-trisected; terminal leaf lobules linear, short-cuspidate; anthodium single on short erect shoots .... 4. B. mongolorum Grub.

+ Small shrubs or subshrubs 5–30 cm tall, with highly lignifying shoots; leaf blade 3–5-palmatisected or simple-pinnatisected; leaf lobules short-cuspidate; anthodia form loose corymbiform or corymbose-paniculate inflorescences, rarely single .... 2.

2. Anthodium without ray flowers, oblong-ovoid, 4–6 mm in diam.; phyllary coarsely foveate-glandular on back ....
.... 1. B. gobicum Krasch.

+ Anthodium with ray flowers, broad-ovoid, 6–10 mm in diam.; phyllary without glandules or with faintly manifest fine glandules .... 3.

3. Leaf blade simple-pinnatisected, with 2–4 pairs of spaced lobules; anthodium single, on long erect pedicels; phyllary with broad membranous margin .... 2. B. kirghisorum Krasch.

+ Leaf blade 3–5-palmatisected; anthodia on short pedicels, usually form loose corymbiform inflorescences at tip of shoots .... 4.

4. Anthodium with short broad-ovoid ray flowers; phyllary without glandules; terminal leaf lobules cuspidate at tip ....
.... 3. B. mongolicum Krasch.

+ Anthodium with long, oblong ray flowers; phyllary thickly strewn with foveate glandules; terminal leaf lobules obtuse at tip .... 5. B. nanschanicum Krasch.

1. B. gobicum Krasch. in Acta Inst. Bot. Ac. Sci. URSS, 1, 1 (1933) 177; Grub. Opred. rast. Mong. [Key to Plants of Mongolia] (1982) 244; Fl. Intramong. ed. 2, 4 (1993) 565; Gub. Konsp. fl. Vneshn. Mong. [Conspectus of Flora of Outer Mongolia] (1996) 100. — Ic.: Krasch. l.c. Plate 198, fig. 8; Grub. l.c. Plate 80, fig. 594; Fl. Intramong. l.c. tab. 223, fig. 5–7.

Described from Mongolia (Gobi Altay). Type in St.–Petersburg (LE). Map 1.

In sandy shrubby deserts, sparse sand, pebble bed-sand floors of gorges, on slopes of red sandstones. Desert species.

IA. Mongolia: *Gobi Alt.* (Bain-Dzak area, 30 km north of Barun-Saikhan mountains, sand in saxaul forest, Sept. 25, 1931-Ik.−Gal., typus!; south. trail of Khurkhe mountains, nor.-east of Shilt-ul mountain, loessial outliers, July 15, 1974−Rachk. et Volk.; 60 km south-west of Nomgon somon, shrubby-sandy desert, July 31, 1978−Volk.), *East. Gobi* (15 km south-west of Shine-usu khuduk, shrubby desert steppe, Sept. 19; 25 km south−south-west of Ulegei khid, thin sand, Sept. 19, 1940; 10 km south-west of Eligen-Tobtsog khuduk on Sain-Shanda−Ulegei khid road, rubbly sand, June 7, 1941−Yun.; old Ulan-Bator−Dalan-Dzadagad road, 2 km north of Talain bulak, feather grass−tansy steppe, Oct. 20, 1947−Grub. et Kal.; Mandal-Obo, old road on Dalan-Dzadagad, Bain-Dzak area, on slopes of red sandstones, July 17, 1948−Grub.; 90 km south of Sain-Shanda−Solong-Khere road, sandy plain, Aug. 30, 1959−Ivan.; Bayan-Dzak ula, on precipices, Sept. 7; Mandal-Obo, 40–45 km south of Khushu khid, Sept. 5–1950, Lavr., Kal., Yun.; 15–20 km nor. of Mandal-Obo somon, right bank of Ongin-gol, steppified shrubby desert, Aug. 12, 1962−Yun. et Dashnyam; Bulgan somon, Bayan-Dzak area nor.-east of somon centre, pebble bed-sandy floor of ravine, July 8, 1970−Sanczir; 45 km south of Ulgii somon, depression between ridges, July 31, 1971−Isach. et Rachk.; Bayan-Dzak area 17 km nor.-west of Bulgan somon centre, red sandstones, Sept. 13, 1979−Grub. et Muld.; Galbyn-Gobi, at boundary in Khuts-ula mountain region, sandy-rubbly slopes of ridges, Aug. 3; north of Khubsugul somon along telegraph line, sandy desert, Aug. 6–1989, Grub., Gub. et Dar.); *Alash. Gobi* (Fl. Intramong. l.c., Gub. l.c.).

General distribution: endemic in Cent. Asia.

2. **B. kirghisorum** Krasch. in Not. syst. (Leningrad), 9 (1946) 171; Tzvel. in Fl. SSSR, 26 (1961) 396; Fl. Kirgiz. 11 (1965) 146; Fl. Kazakhst. 9 (1966) 69; Opred. rast. Sr. Azii [Key to Plants of Mid. Asia] 10 (1993) 589.−*B. pulvinatum* (Hand.-Mazz.) Shin var. *kirghisorum* (Krasch.) in Fl. Sin. 76, 1 (1983) 28.−*Chrysanthemum fruticulosum* auct. non DC.: O. et B. Fedtsch. Consp. Fl. Turk. 4 (1911) 183. −Ic.: Krasch. l.c. Plate I, Fl. Kazakhst. l.c. Plate 8, fig. 8.

Described from Kirghizia (Cent. Tien Shan). Type in St.-Petersburg (LE).

On rocky, rubbly slopes, rocks marmorized limestones, in arid river valleys.

IB. Kashgar: *South.* (60 km south−south-west of Kerii settlement, on road to Polur, nor. slope of Kun-Lun, along slope of gorge, 2200 m, May 10; nor. slope of Kun-Lun, 7–8 km north of Polur, on road to Kerii, steep sand-covered slope, 2500 m, May 13–1959, Yun.), *West.* (4 km south of Ak-Kez-daban pass along Tibetan highway, on limestones in mountainous desert, June 4; 3–4 km west of Ak-Kez-Daban pass along highway to West. Tibet from Kargalyk, arid right-bank valley entering Tiznaf, on slopes of marmorized limestones in mountainous saltwort desert, June 5–1959, Yun.).

General distribution: endemic in Cent. Asia.

36   Note. The reference cited in Fl. Sin. l.c. to the find of *B. fruticulosum* (Ledeb.) DC. in Sinkiang territory has not been confirmed by our herbarium material and calls for verification.

3. B. mongolicum Krasch. in Not. syst. (Leningrad), 11 (1949) 196; Grub. Opred. rast. Mong. [Key to Plants of Mongolia] (1982) 244; Fl. Sin. 76, 1 (1983) 26; Gub. Konsp. fl. Vneshn. Mong. [Conspectus of Flora of Outer Mongolia] (1996) 100. —Ic.: Krasch. l.c. Plate 195, fig. 6.

Described from Mongolia (Jung. Gobi). Type in St.-Petersburg (LE).

In petrophyte associations, on granite outliers, desert slopes of hummocky areas.

IIA. Junggar: *Jung. Gobi* (nor. slope of Tien Shan, at Pichan-Kichi-Ulansu meridian, June 1889—Gr.-Grzh., paratypus!; pass in Dzhirgalanta valley, Sept. 16, 1930—Bar., typus!; Oshigiin-usu area, along granitic hummocky area, July 30, 1947—Yun.; 7 km south-east of Upkeituk-ula, winter fat—saxaul desert, Sept. 19, 1948—Grub.; Dashitu area, Muleikhe district, on slope, Sept. 24, 1957—Yun.; 60 km south of Ertai settlement (on Urungu) on road to Guchen, barren hummocky area, July 16, 1959—Yun. et Yuan'; 75 km south-west of Bulgan somon near Ushigiin-ula spring, slopes of granitic plateau, Aug. 7; 5 km from Uench somon, along trails of conical hillock in winter fat barren land, Aug. 17–1977, Volk. et Rachk.; vicinity of Bulgan somon, Dzagyn-Ulan low mountain, petrophyte associations, July 17; vicinity of Uench somon, petrophyte associations, July 27–1984, Dar., Kam.).

General distribution: endemic in Cent. Asia.

4. B. mongolorum Grub. in Bot. zhurn. 57, 12 (1972) 1593; Grub. Opred. rast. Mong. [Key to Plants of Mongolia] (1982) 244; Gub. Konsp. fl. Vneshn. Mong. [Conspectus of Flora of Outer Mongolia] (1996) 100. —Ic.: Grub. l.c. Plate 80, fig. 595.

Described from Mongolia (East. Mong.). Type in St.-Petersburg (LE).

In petrophyte associations on rocky summits and crests of conical hillocks, on rocks.

IA. Mongolia: *East. Mong.* (16 km nor.-east of Aradzhargalantykhid, west. part of Tsogt-Undyr mountains, petrophyte associations on rocky summits, June 23, 1971—Dashnyam, Isach., Karam., Rachk., Safronova, typus!; 45 km south-west of Sukhe-Bator, along out-crops of limestones on summits of conical hillock, July 5, 1971—Isach., Rachk.; 29 km south of Barun-Urt, petrophyte associations along rocky summits, July 10, 1971—Dashnyam, Karam., Safronova; 15 km west of Barun-Matad-ul centre, along mountain crests, on rocks, Aug. 12, 1989—Sanczir et Khramtsov).

General distribution: endemic.

5. B. nanschanicum Krasch. in Not. syst. (Leningrad), 11 (1949) 200.— *B. pulvinatum* auct. non (Hand.-Mazz.) Shin: Fl. Sin. 76, 1 (1983) 27; Fl. Intramong., ed. 2, 4 (1993) 565. —Ic.: Krasch. l.c. Plate 196, fig. 7.

Described from Qinghai (Qaidam). Type in St.-Petersburg (LE). Plate III, fig. 3. Map 1.

In desert associations on clayey soils, rocks, pebble bed terraces, on arid river beds.

IA. Mongolia: *Alash. Gobi* (Alashan south., Aug. 15, 1880—Przew.; Bayan-Khoto, Tengeri sand, Taogien'—Shan', June 31, 1958—Petr.), *Khesi* (between Baiduntsza and Sayan'tszyn villages, Sept. 20, 1901—Lad.; 30 km east of An'si town on road to Yuimyn', south. extremity of Bei-Shan', along ravines, Oct. 7, 1957 — Yun. et Lee; 60 km west of Tszyutsyuan', eroded loamy sand submontane plain, July 24; 55 km east of Chzhan'e town, hillocky Nanshan foothills, Aug. 10-1958, Petr.; vicinity of Tszyutsyuan' town, Tszya-Yui-Guen' old fort, desert associations along slopes and pebble bed terraces, Sept. 21; Chi-lan' -shan' mountains (west. part of Richthofen mountain range), Tszyutsyuan' town — Yuimyn' town, on loamy sandy soil in saltwort desert, Sept. 22-1958, Lavr.).

37   IC. Qaidam: *mount.* (Puteum Mantschzu-bulak, Aug. 23, 1879 — Przew., typus!; 15 km south of Aksai settlement, Altyntag mountain range, gorge slope, 2800 m, Aug. 2, 1958 — Petr.).

IIIA. Qinghai: *Nanshan* (nor. slope of Nanshan, 1875 — Pias.; nor. slope of Humboldt mountain range, June 24, 1895 — Rob.).

General distribution: endemic to Cent. Asia.

Note. Fl. Sin. l.c. and Fl. Intramong. l.c. cite species *B. pulvinatum* (Hand.-Mazz.) Shin with 2 varieties *B. pulvinatum* var. *kirghisorum* (Krasch.) and *B. pulvinatum* var. *fruticulosum* (Ledeb.) DC. for Qaidam, Qinghai and entire Sinkiang territories. Our study of type and herbarium material shows that 2 species are found here. *B. tianschanicum* Krasch. described from this territory differs greatly from *B. pulvinatum* in several morphological characteristics: for example, ray flowers in *B. pulvinatum* are oval and phyllary smooth-edged; in *B. nanschanicum*, these flowers are narrow, oblong and phyllary with lacerated-undulated margin.

Moreover, after I.M. Krascheninnikov, we treat *B. kirghisorum* Krasch., fairly extensively distributed in the mountainous deserts of Kashgar, as an independent species.

In Sinkiang territory, according to our data, only *B. mongolicum* Krasch. and not *B. fruticulosum* (Ledeb.) DC. is found in Sinkiang territory as pointed out in Fl. Sin. l.c.

14. Phaeostigma Muld.
Bot. zhurn. 66, 4 (1981) 586

1. Ph. salicifolium (Mattf.) Muld. in Bot. zhurn. 66, 4 (1981) 587. — *Tanacetum salicifolium* Mattf. in J. Arn. Arb. 13 (1932) 407. — *Chrysanthemum linariifolium* Shang, in Sinensia, 5 (1934) 160. — *Ch. salicifolium* (Mattf.) Hand.-Mazz. in Acta Horti Gotob. 12 (1938) 264; Hu in Quart. J. Taiwan Mus. 17 (1966) 41. — *Ajania salicifolia* (Mattf.) Poljak. in Not. syst. (Leningrad) 17 (1955) 424. — Ic.: Poljak. l.c., fig. 6 (a–i); Iconogr. Cormophyt. Sin. 4 (1975) 517, No. 6448.

Described from China (Gansu). Type in London (K). Isotype in St.-Petersburg (LE). Plate IV, fig. 4.

In shrubby thickets, meadowy slopes of mountains, 2100–3100 m alt.

IIIA. Qinghai: *Nanshan* (Rangkhta-gol river, July 24, 1872 — Przew., isotypus!, Yusun-Khatyma river, Aug. 11; South Tetung mountain range, 2600 m, Aug. 20-1877, Przew.; Gan'chan-Gomba, on meadowy slopes of mountains, 2800-3100 m, Sept.; Tetung river, along banks of Dai-Tung-kho river, in scrubs, 2100 m, Sept. 2, 1901, Lad.; "Datun"; "Min'khe" — Fl. Sin. l.c.).

General distribution: China (Gansu, Sichuan).

15. Ajania Poljak.

Not. syst. (Leningrad), 17 (1955) 419. — *Cryanthemum* R. Kam. Opred. rast. Sr. Azii [Key to Plants of Mid. Asia] 10 (1993) 635.

1. Small subshrubs up to 20 cm tall, with highly stunted perennial vegetative shoots and smooth-edged lobed leaves; anthodium terete, phyllary imbricate .... 15. A. trifida (Turcz.) Tzvel.

+ Subshrubs or small subshrubs with well-developed caudex; lower and middle cauline leaves pinnatisected or incised into 3–9 first order lobes, in turn wholly or partly 3–5-incised or lobed; anthodium of different form .... 2.

2. Plant 15–60 (90) cm tall; anthodium relatively small, 2.5–4.5 mm in diam., form compound corymbose inflorescence; achene slimy on wetting .... 3.

+ Plants 2–15 (20) cm tall; anthodium relatively large, 5–10 mm in diam., 2–8 (10) of them form compact or loose corymbose inflorescence at end of fertile shoot; achene not slimy on wetting .... 14.

3. Terminal leaf lobules of lower and middle cauline leaves filiform-linear, up to 0.5 mm broad; plant pubescent with double-ended and simple long hairs ....
.... 8. A. nematoloba (Hand.-Mazz.) Ling et Shih.

+ Terminal leaf lobules of lower and middle cauline leaves of different form and usually broader than 0.5 mm; plant pubescent with only double-ended hairs .... 4.

4. Lower and middle cauline leaves long-petiolate, considerably longer than leaf blade; anthodia campanulate, form compact corymbose inflorescence; outer phyllary ovoid ....
.... 12. A. roborowskii Muld.

+ Lower and middle cauline leaves on petioles as long as leaf blade or shorter; anthodium of different form; outer phyllary linear to broad-lanceolate .... 5.

5. Leaves pinnatipartite but appear 3–5-palmate due to highly proximated lobes .... 6. A. grubovii Muld.

+ Leaves of different structure, terminal leaf lobules more apart .... 6.

6. Leaf blade greyish on both sides due to pubescence; terminal leaf lobules obovoid or narrow-obovoid; carina and pubescence on phyllary distinctly manifest ....
.... 1. A. achilleoides (Turcz.) Poljak. ex Grub.

+   Leaf blade greyish green above, greyish underneath due to compact fine-cobwebby pubescence; terminal leaf lobules linear or linear lanceolate .... 7.

7.   Fertile shoots highly branched at base and woody to quite a height; blades of lower and middle cauline leaves 2-pinnatisected or 3-palmatipartite .... 8.

+   Fertile shoots not branched at base; blades of lower and middle cauline leaves 2-pinnatipartite with 5–7 first order lobes ....
.... 3. A. fastigiata (Winkl.) Poljak.

8.   Leaf blade of lower and middle cauline leaves pinnatisected; terminal leaf lobules lanceolate or linear-lanceolate .... 9.

+   Leaf blade of lower and middle cauline leaves 3-palmatipartite; terminal leaf lobules narrow-linear or linear .... 11.

9.   Fertile shoots erect; anthodium on long pedicels, form compact corymbose inflorescence; outer phyllary ovoid, sparsely pubescent .... 10. A. przewalskii Poljak.

+   Fertile shoots ascending at base, brown or violet-brown; anthodium on short pedicels form loose corymbose inflorescence; outer phyllary compactly pubescent .... 10.

10.   Terminal leaf lobules of lower and middle cauline leaves oblong to broad-linear, subobtuse at tip; phyllary in 2–3 irregular rows .... 5. A. gracilis (Hook. f. et Thoms.) Poljak. ex Tzvel.

+   Terminal leaf lobules of lower and middle cauline leaves narrow-linear, subacute at tip; phyllary in 3–4 irregular rows ....
.... 4. A. fruticulosa (Ledeb.) Poljak.

11 (8).   Small subshrubs with rather few dark brown, rather sparsely leafy erect fertile shoots; blades of lower and middle cauline leaves short-petiolate or subsessile .... 12.

+   Small subshrubs with many light brown compactly leafy fertile shoots ascending at base; blades of lower and middle cauline leaves on stalks as long as leaf blade or slightly longer .... 14.

12.   Terminal leaf lobules narrow-lanceolate, 3–5 mm long, subobtuse at tip; anthodia narrow-ovoid, form fairly compact corymbose inflorescence .... 9. A. parviflora (Gruning) Ling.

+   Terminal leaf lobules lanceolate or broad-lanceolate; anthodia ovoid or broad-ovoid, form loose inflorescence or 2–3 of them together at tip of shoots .... 13.

13.   Terminal leaf lobules lanceolate, 3–5 mm long, obtuse at tip; anthodia oblong-ovoid, form loose corymbose inflorescence ......
.... 11. A. purpurea Shih.

+ Terminal leaf lobules broad-lanceolate, up to 3 mm long, short-cuspidate at tip; anthodium ovoid, 2–3 of them together at tip of fertile shoots .... 2. A. alabasica H.C. Fu.

14. Plant 5–20 cm tall with poorly developed caudex; blades of lower and middle cauline leaves trisected; terminal leaf lobules entire, rarely all or some of them 2–5-partite; anthodia in compact corymbose inflorescence .... 15.

+ Plant 2–10 (12) cm tall with highly branched caudex; blades of lower and middle cauline leaves 2-pinnatisected; anthodium single or in loose corymbose inflorescences .... 16.

15. Lower and middle cauline leaves on petioles as long as leaf blade; anthodium on pedicels with white tomentose pubescence, 3–8 (10) of them on single shoot; phyllary with white tomentose pubescence at base ....
.... 16. A. trilobata Poljak. ex Tzvel.

+ Lower and middle cauline leaves on vary short petioles, subsessile; anthodia 2–3 (5) together on a single shoot; phyllary without tomentose pubescence at base ....
.... 7. A. khantensis (Dunn) Shih.

16. Terminal leaf lobules of lower and middle cauline leaves obovoid, 1–1.5 mm long, 1–1.2 mm broad, obtuse at tip ....
.... 14. A. tibetica (Hook. f. et Thoms.) Tzvel.

40 + Terminal leaf lobules of lower and middle cauline leaves broad-linear or lanceolate, subobtuse, 2–4 mm long, up to 1 mm broad .... 13. A. scharnhorstii (Rgl. et Schmalh.) Tzvel.

1. A. achilleoides (Turcz.) Poljak. ex Grub. in Novit. syst. pl. vasc. 9 (1972) 296; Grub. Opred. rast. Mong. [Key to Plants of Mongolia] (1982) 245; Fl. Sin. 76, 1 (1983) 122; Fl. Intramong. ed. 2, 4 (1993) 589; Gub. Konsp. fl. Vneshn. Mong. [Conspectus of Flora of Outer Mongolia] (1996) 96.—*Artemisia achilleoides* Turcz. in Bull. Soc. natur. Moscou, 5 (1832) 193; Maxim. in Mel. Biol. 11 (1872) 520.—*Tanacetum achilleoides* (Turcz.) Hand.-Mazz. in Acta Horti Gothob. 12 (1938) 27; S.J. Hu in Quart. J. Taiwan Mus. 19 (1996) 24.—*Hippolytia achilleoides* (Turcz.) Poljak. ex Grub. Konsp. fl. MNR [Conspectus of Flora of Mongolian People's Republic] (1955) 262. —Ic.: Fl. Sin. l.c. tab. 20, fig. 1; Fl. Intramong. l.c. tab. 234, fig. 1–4.

Described from Mongolia (East. Gobi). Type in Kiev (KW). Isotype in St.-Petersburg (LE). Map 2.

Desert-steppe species distributed on rocky and stony slopes of low mountains among feather grass—cereal grass, biurgun (*Anabasis brevifolia*)—feather grass desert-steppe associations with shrubs, as well as along floors of gorges and rock crevices.

IA. Mongolia: *Mong. Alt.* (Khara-Adzarga mountain range, valley of Sakhir-sala river, along coastal pebble bed, Aug. 21; Khasagtu-Khairkhan, entrance to Dundu-seren-gol, rocky trail, Sept. 15–1930, Pob.; nor. slope of Mong. Altay, Khalyun area, rocky slope, Aug. 24, 1943 – Yun.), *Cent. Khalkha* (Mogoitu field camp [250 km south-east of Ulan-Bator town], along slopes of mountains, Aug. 10, 1831, No. 203 – I. Kuznetsov, isotypus!; Khodotu station, on rocky soil, Aug. 10, 1831, No. 233 – Ladyzh.; Uber-Dzhergalante river, near Dol'che-Gegen monastery, rubbly steppe, Aug. 20, 1925 – Krasch. et Zam.; 75 km south of Choiren, on slopes, Aug. 24, 1926 – Lis.; on granitic slopes of Bichikte massif, Aug. 6, 1928 – Shastin; along road 6 km south of field camp of Del'ger-Tsogtu somon, rubbly summit of granitic conical hillock, Aug. 12, 1950 – Lavr.), *Depr. Lakes* (Shargain-Gobi desert, on way from Barun-khuduk collective to Gol-Ikhe, rocky trail, Sept. 14, 1930 – Pob.; nor. bank of Khirgis lake, on south. slope of Khan-Khukhei mountain range, Aug. 17, 1977 – Karam., Bonzragch et al; west. fringe of Shargain-Gobi on road to Altan-Shire somon, 19 km from centre of ajmaq [administrative territorial unit in Mongolia], exposed red-coloured formations, snakeweed steppe, Aug. 9, 1979 – Grub., Muld., Dar.), *Gobi Alt.* (south. slope of Dundu-Saikhan mountains, July 13, 1909 – Czet.; foothills of Ikhe-Bogdo mountain, eroded moraine, Aug. 24, 1926 – Tug.; Dundu-Saikhan mountains, in rock crevices, Aug. 17; Dzolen mountains, on rocks along slope, Sept. 8; Dzun-Saikhan mountains, in rock crevices, Sept. 24–1931, Ik.-Gal.; Gurban-Saikhan mountain range, rocky slope, June–July 1939 – Donoi Surmazhab; Ikhe-Bogdo mountain range, nor. slope of midportion of Bityuten-ama creek valley, juniper groves on rocky slopes, Sept. 12; Noyan-Bogdo-ula, 4–5 km south of Noyan somon, feather grass desert steppe on rocky slope, July 25–1943, Yun.; Nomogon somon, Khurkhe mountain range, rocky slope of Altyn-ala gorge, Sept. 7, 1950 – Kal. et Yun.; Barun-Saikhan mountain range, mountain summit, Aug. 23, 1953 – Dashnyam; 22 km west – south-west of Gurban Tes settlement, Tasty-nuru mountain range, south. rocky slope, July 27, 1972 – Rachk. et Guricheva; east. extremity of Dzolen mountain range, southward of Ulan-Des-ul on road to Noen somon, saltwort – biurgun (*Anabasis brevifolia*) semidesert, Sept. 9; same site, Dzun-Saikhan mountain range, Khuren-Dzag mountains, 15 km from Dalan-Dzadagad on road to Khurmein somon, along floor of ravine, Sept. 12; khuren-khana mountain range, along Khairein-Khundei gorge near Elistiin-Dzakhoi spring, 1600 m, in rock crevices, Sept. 6–1979, Grub., Muld., Dar.), *East. Gobi* (47 km south-east of Choiren on road to Sain-Shandu, feather grass desert steppe, Aug. 26; Ulegei-khid, 1–2 km west of monastery, rocky summit of conical hillock, Sept. 20–1940, Yun.; 125 km from Sain-Shanda on road to Ulan-Bator, rocky plain, Sept. 5, 1950 – Ivan.; 60 km south – south-east of Khubsugul somon, Khutag-ula mountain, steppified throny cushion formation, Aug. 28; 50 km south-east of Ulgii, along slope of outlier, July 31; same site, 27 km south of patrol No. 25, rocky summit of conical hillock, June 14–1971, Isach. et Rachk.; 50 km north-east of Mandal-Gobi somon, hummocky area, July 14; 60 km east – south-east of Mandal-Gobi somon, hummocky massif toward north-west of Buyant-ul mountains, on summit of conical hillock, June 14; 20 km east of Denger-Khangai somon, on south. slope of conical hillock, June 14; south-east of Mandal-Gobi, on south. slopes of hummocky area, July 17; 80 km south-east of Mandal-Gobi ajmaq [administrative territorial unit in Mongolia], 2 km west of old Gurban-Saikhan somon, feather grass steppe, July 17-Rachk.; 58 km from Erdeni-Dalai somon, Sept. 11, 1974 – Sanzhid), *Val. Lakes* (Khailyaste creek valley near Lamyn-Khita on Argiin-gol river, Aug. 21, 1926 – Lis.; 30 km west of Dzhiin setu somon, steppe with subshrubs, Aug. 17 – Kal.), *Alash. Gobi* (Bordzon-Gobi, south. slope north-east of Khaldzan-ul mountain range, Sept. 8, 1950 – Lavr., Kal., Yun.; west. extremity of south. slope of Khaldzan-ul mountain range, on rocks, July 3, 1989 – Grub., Gub., Dar. et al).

General distribution: endemic in Cent. Asia.

2. A. alabasica H.C. Fu in Fl. Intramong. 6 (1982) 34, 325; Fl. Intramong., ed. 2, 4 (1993) 585. — Ic.: Fl. Intramong. 4 (1993), l.c. tab. 232, fig. 1–5.

Described from Inner Mongolia (Ulantsab). Type in Khukh-Khoto (HIMC).

On rocky slopes.

IA. Mongolia: *East. Gobi* ("Ulantsab" — Fl. Intramong. 6, l.c.).

General distribution: endemic in Cent. Asia.

3. A. fastigiata (Winkl.) Poljak. in Not. syst. (Leningrad), 17 (1955) 428; Tzvel. in Fl. SSSR, 26 (1961) 405; Fl. Kirgiz. 11 (1965) 149; Fl. Kazakhst. 9 (1966) 71; Fl. Sin. 76, 1 (1983) 125; Opred. rast. Sr. Azii [Key to Plants of Mid. Asia] 10 (1993) 590. — *Artemisia fastigiata* Winkl. in Acta Horti Petrop., 11, 2 (1891) 373. — *Tanacetum fruticulosum* Ledeb. Fl. Ross. 2, 6 (1845) 603, quoad. pl. ex Alatau Dshung. — Ic.: Fl. Kazakhst. l.c. Plate 9, fig. 1.

Described from Mid. Asia (region between Alay and Peter I mountain ranges). Type in St.-Petersburg (LE).

On rocky desert-steppe and shrubby steppe slopes of mountains, 1400-3000 m abs. alt.

IB. Kashgar: *East.* (Kyzyltag mountain range, 25 km south-west of Kyumysh, mountain desert steppe, Sept. 9, 1959 — Petr.).

IIA. Junggar: *Dzhark.* (Pilyuchi river, June 21, 1878–Fet.); *Tien Shan* (valley of Ili river, Sept. 29, 1876–Przew.; Maralty, near Muzart entrance to Tekes river valley, Aug. 1, 1877 — Fet.; Talki gorge, July 23; Sairam, July 26; Chapchagai, Aug. 9–1877, A. Reg.; Kash river, between Ulyastai and Nilki, 2000 m, June 30; Monguto, about 3000 m, Aug.; Kunges river, upper course of Algoi river, Sept. 12–1879, A. Reg.; Ketmen' mountain range, valley of Sarbushin river near its emergence from mountains onto trail, south. rocky slope, Aug. 2; valley of Tekes river, 7 km south-east of Aksu settlement, mountain steppe with shrubs, Aug. 24–1957, Yun. et Lee; 25 km north of Tsitszyaotszin', along floor of arid gorge, Oct. 3, 1959 — Petr.).

General distribution: Jung.-Tarb., Nor. and Cent. Tien Shan, Mid. Asia (west. Tien Shan, Pam.-Alay).

4. A. fruticulosa (Ledeb.) Poljak. in Not. syst. (Leningrad) 17 (1955) 428; Tzvel. in Fl. SSSR, 26 (1961) 406; Fl. Kazakhst. 9 (1966) 71; Grub. Opred. rast. Mong. [Key to Plants of Mongolia] (1982) 245; Fl. Sin. 76, 1 (1983) 123; Fl. Xizang. 4 (1985) 738; [Drevesn. rast. Tsinkhaya-Wooded Plants of Qinghai] (1987) 629; Fl. Intramong. ed. 2, 4 (1993) 590; Opred. rast. Sr. Azii [Key to Plants of Mid. Asia] 10 (1993) 590; Gub. Konsp. Fl. Vneshn. Mong. [Conspectus of Flora of Outer Mongolia] (1996) 96. — *Tanacetum fruticulosum* Ledeb. Fl. Ross. 2, 6 (1845) 603; Kryl. Fl. Zap. Sib. 11 (1949) 2758. — *Artemisia athanasia* Bess. in Bull. Soc. natur. Moscou, 7 (1834) 24. — Ic.: Ledeb. Ic. pl. Fl. Ross. 1 (1829) 10, tab. 38; Fl. Kazakhst. 9, Plate 9, fig. 2; [Drevesn. rast. Tsinkhaya-Wooded Vegetation of Qinghai] l.c. tab. 448; Fl. Intramong. l.c. tab. 234, fig. 5–9.

Described from Altay. Type in St.-Petersburg (LE).

Desert-steppe species distributed on rocky and rubbly slopes of low mountains to the midmountain belt, comprises feather grass—tansy, saltwort—wormwood, biurgun (*Anabasis brevifolia*)—feather grass with *Reaumuria* associations, rarely found in saxaul barren lands, along gorges.

IA. Mongolia: *Khobd.* (east. portion of Achitnur basin, biurgun-feather grass steppe, June; Sagliin somon, south-western extremity of Uryuk-nur basin, before ascent to pass, feather grass barren steppe rocks, July 30; midcourse of Tsagan-Nuriin-gol, Obgor-ul, in lower mountain belt, Aug. 2–1945, Yun.), *Mong. Alt.* (vicinity of Tsakhir-Bulak spring, on arid river bed, Aug. 18; Taishiri-ula mountain range, Aug. 18–1894, Klem.; Bombotu-Khairkhan mountains, talus along mountain slopes in Khapchik river valley, Oct. 10, 1930—Pob.; granitic conical hillock on right bank of Buyanty-gol, 2 km north of Khobdo, on slopes, oct. 2, 1948—Grub.), *Depr. Lakes* (Daribi somon, south-eastern extremity of Khoisiin-gobi, 5–7 km east of Bain-gol workshop, biurgun-wormwood—feather grass steppe, Aug. 24, 1944—Yun.; Mankhan somon, Bodkhon-gol area, 35 km south of Tugrik-Sume, feather grass barren steppe, Aug. 11; Tonkhil somon, east of slope toward Tonkhidnur basin, feather grass, mountainous barren steppe, Aug. 13–1945, Yun.; 2 km west of Sundultu-Baishing, along banks and floor of arid irrigation canal, Sept. 4, 1948—Grub.), *Val. Lakes* (not far from Orok-nor lake, July 13, 1893—Klem.; Orok-nor, Naryn-Khara mountains, Aug. 12, 1926—Tug.; between Tuin and Tatsiin rivers, basalt rocks and placers, Aug. 15, 1943—Kal.; east. vicinity of Guiliin-Tal area, 10 km west of Bain-Khongor somon, feather grass steppe, Aug. 27, 1943; gently inclined nor. slope of Margats-ula, saltwort—feather grass barren steppe, Aug. 19, 1944—Yun.; 20 km nor.—nor.-west of Delger somon on roadside, feather grass steppe along fringe of hummocky area, Aug. 30, 1948—Grub.; Dzhinsetu somon, Barun-Khongor area, steppe on sand-covered soil, Aug. 17, 1949—Kal.; 6 km west of Dzapkhan somon, feather grass-semishrub subdesert, 1100 m, Aug. 9, 1979—Rachk. et Sumerina), *Gobi Alt.* (nor. slope of Tostu mountain range, Aug. 19, 1880—Pot.; southward of Baga-Bogdo, in ravine, Aug. 9, 1894—Klem.; nor. slope of Dzolen mountain (its nor. trail) covered with thin sand, July-Aug.; valley of Legin-gol river, barren hillocks, Aug. 19–1927, Simukova; Bain-Tukhum area, on rubble trail of Bain-Tsagan mountain, Aug. 30, 1931—Ik.-Gal.; Bain-Tsagan mountain range, slopes from trail to upper mountain belt, Aug. 7, 1933—Khurlat et Simukova; Dzun-Saikhan mountains, 6 km south of Dalan-Dzadagad, rocky steppe, July-Aug.—Donoi Surmazhab; Noyan-Bogdo mountains, 2 km south of Noyan somon, feather grass barren steppe, July 25; nor. trail of Chindamani-ul mountain (its midportion), wormwood—feather grass barren steppe, Aug. 22; Ikhe-Bogdo mountain range, nor. slope of Bityuten-ama creek valley, along floor of gorge, Sept. 12–1943, Yun.; Baga-Bogdo mountain range, halfway from Dzhargalant-khuduk well and Ikhe-Bogdo-ula, saltwort—feather grass barren land, July 20; 25 km north of Nemegetu-ul, in Barun-Bogdo-obo mountains and Khara-Tologoi-khuduk collective, poplar-saxaul oasis, Aug. 3; nor. trail of Tostu-nuru mountain range, between Oidol-khuduk and Khushu-khuduk wells, 7 km from former, biurgun barren land, Aug. 9; Noyan somon, south. bend on road to Chonoin-Bom area, rocky hummocky area, among rocks, Aug. 18; south. gently inclined trail of Gobi Altay 20 km nor.-west of Dzadagai-Khere, feather grass-biurgun with *Reaumuria* steppe, Aug. 25–1948, Grub.; Arts-Bogdo mountain range, nor. slope, 98 km nor.-west of Bulgan somon, floor of gorge, July 11, 1970—Karam., Lavr., Bonzragch; 40 km south of Dalan-Dzadagad, winter fat—*Ayenia*—feather grass steppe, July 14, 1974—Rachk. et Volk.), *East. Gobi* (Baga-ude, sand gully near Urguni collective, Aug. 15, 1926—Polynov et Lis.; on road to Yaman-Ikhe-Dulan-Khoshun, Aug. 20, 1928—Shastin; Khan-Bogdo somon, Khoir-Ul'tszeitu area, near Khutag-ula, barren steppe,

43   Aug. 23; Delger-Khangai somon, Delger-Khangai mountain range, rocky slope of mountain, Sept. 8; Kholtu somon, southward of Sharan-Gutai workshop, Sept. 15–1930, Kuznetsov; Motonge mountains, 40 km north of Dzamygin-ude, sandy valley and rocky slope, Aug. 30, 1931 – Pob.; 75 km nor.-east of Dalan-Dzadagad on road to Ulan-Bator, barren steppe, Aug. 16; 11 km south-east of Talain-bulak on road to Bain-Dzag from Dalan-Dzadagad, feather grass barren steppe, Sept. 11; Altyn-Shire somon, 72 km nor.-east of Sain-Shanda on road to Baishintu, feather grass barren steppe, Sept. 13; 20 km south-west of Shine-usu-khuduk on road to Ulegei-khid, wormwood desert, Sept. 19; nor. foothill of Khan-ul mountain, near Dalan-Dzadagad-Tsogtu – Tsetsei somon road, feather grass desert steppe, Oct. 5; 10–12 km east of Argalinte somon on road to Sudzhi-khuduk, feather grass barren steppe, Sept. 2–1940, Yun.; Mandal-obo somon, old Ulan-Bator-Dalan-Dzadagad road, 2 km northward, feather grass-tansy steppe on somewhat ridgy plain, Oct. 20, 1947 – Grub. et Kal.; Godli-Balgasuin-Tala intermontane plain, southward of Khan-ul mountain range, alliaceous – feather grass steppe, Sept. 7, 1950 – Lavr.; Bain-Dalai somon north of Dalan-Dzadagad, Sept. 15, 1951 – Kal.; Bayan basin 80 km west – nor.-west of new Bayan-obo somon (old Dzhargalantan) on road to Dalan-Dzadagad, flat submontane plain, June 27; 145 km south-east of Mandal-Gobi ajmaq – [administrative territorial unit in Mongolia]–pea shrub – wormwood feather grass steppe, July 19–1972, Rachk. et Guricheva; Bayan-Dzag basin, 18 km north-east of Bulgan somon, July 13, 1972 – T. Popova; 26 km north of Bulgan somon, Aug. 4, 1972 – Isach.; 60 km east-south – east of Khan-Bogdo somon on road to Khatan-Bulak somon, in gorge, July 20, 1974 – Rachk. et Volk.; 10 km north of Bulgan somon, *Ajania* barren steppe, June 8, 1975 – Kazantseva; 1 km south – south-west of Bulgan somon, Gurban-Saikhan submontane plain, shrubby steppe, Aug. 2, 1976–Bespalova), *West. Gobi* (Dzakhoi-Dzyuram area, barren steppe, Aug. 20, 1943 – Yun., south. foothills of Tsagan-Bogdo mountain range, rocky hummocky area, Aug. 1; 1–2 km east of Bortsu-bulak east of Atas-Bogdo mountain trail, hummocky area, Aug. 11; Atas-Bogdo-ula, on rocky slope and floor of gorge, Aug. 12; 40–50 km south of Maikhan-bulak, rubbly-rocky barren area with saxaul, Aug. 16; between Atas-Bogdo and Maikhan bulak, saxaul barren land, Aug. 16, 1943; Tukhumyin-Khundei ravine in Khubchiin-nuru mountain range, north-west of Adzhi-Bogdo, barren wormwood – feather grass steppe, Aug. 9, 1947 – Yun.; 1 km north of Burkhantu-bulak spring on road to pass, along ravine, Aug. 23, 1948 – Grub.; Ansi in Shinshinsya district, rocky slopes, Oct. 5, 1959-Petr.; Khoshuin-nuru mountain range, deep gorge, July 26, 1973 – Grub. et al.; 5 km west of Khatyn-Suudal spring, Khatyn-Suudalyn-nuru mountain range, on nor. slope, July 21; 50 km west – north-west of Tsagan-bulak post, on Khatyn-Suudal road, gorge floor, July 21, 1973; Atas-Bogdo mountain range, nor. macroslope, pasture sage brush-wheat grass mountain steppe, Aug. 23, 1976 – Rachk. et Damba), *Alash. Gobi* (Alashan mountain range, Sept. 1871; Alashan-urga, Aug. 1873; South. Alashan, near Dadzhin town, Aug. 14; Alashan mountain range, Aug. 20; south. Alashan, Aug. 20–1880, Przew.; Kobden-usu area, Aug. 13; Kobden-usu area, nor. Gashun-nor lake, Aug. 14; same site, nor. of Gashun-nor lake, Aug. 19; Urzhyum collective region, Aug. 16–1880, Pot.), *Ordos* (near Linchzhou town, Sept. 16; Gaosovo, Sept. 28–1884, Pot.), *Khesi* (vicinity of Lanchzhou, along upper course of Vei-kho river, June 7, 1875 – Pias.; nor. slope of Altyn-tag mountain range, on loessial foothills, Aug. 6, 1890 – Rob.; Beishan' mountains, Lan'chzhou town, Sept. 6; same site, steppe near Khuikhu, Sept. 13, 1890 – Gr.-Grzh.; Suchzhou area, Sept. 22, 1890 – Marten; Tien-Shan-pui mountain range, San'tszin, along arid clayey ridges, Sept. 19, 1901 – Lad.; 100 km south-west of Dunkhuan, arid river bed on alpine nor. foothills of Altyn-tag mountain range, Aug. 21, 1958 – Petr.).

   IB. Kashgar: *South.* (Keriya-Dar'ya, near Lyushi estuary, June 30, 1885 – Przew.), *West.* (Tokhtakhon area, on loessial mountain slope, July 17, 1889 – Rob.), *East.* (Algoi river, Sept. 12; nor.-west of Turfan, Sept. 12, 1879 – A. Reg.).

46

IIA. Junggar: *Jung. Ala Tau* (Dzhair mountain range, Tuz-agny meadow in Turangy-bastau oasis, on clayey terrace, Sept. 8, 1951 – Mois.), *Jung. Gobi* (Khubchiin-nuru hummocky area, feather grass barren steppe, July 3; 20 km east of Oshigiin-usu area, saxaul barren area, in ravine, July 31; 7–8 km east of Guntamchi area, wormwood barren steppe on loamy sand, Aug. 2–1947, Yun.; nor. slope of Baga-Khabtag-nuru mountain range, on rocky slopes, 1800 m, Sept. 14; Baitak-Bogdo-nuru mountain range foothills, 6 km from Utszyur-Tsagan-nuru, *Ephedra*-saltwort barren land along trail, Sept. 15–1948, Grub.; 105 km south of Ertai (in Urumchi) on road to Guchen, rocky saxaul barren area, July 16, 1959 – Yun.; Sertengiin-Khuvch mountain range, 25 km south-east of Altay somon, July 6, 1984 – Kam. et Dar.).

IIIA. Qinghai: *Nanshan* (Ushilin pass, July 21, 1875 – Pias.; Humboldt mountain range, nor. slope toward Chansai gorge, Mogyn-Dybit area, on rocks, June 24, 1895; on loessial arid slopes of Chansai area, July 16, 1889 – Rob.; 15 km south of Aksai settlement, Altyntag mountain range, rocky slopes, Aug. 28, 1958 – Petr.).

IIIB. Tibet: *Chang Tang* ("Zhitu, Getszi"), *South.* ("Chzhada" – Fl. Xizang. l.c.).

IIIC. Pamir (Pas-Rabat post, July 3, 1909 – Divn.).

General distribution: Fore Balkh., Jung.-Tarb., West. and East. Sib., Nor. Mong. (Mong.-Daur.).

Note. An unusually polymorphous species. Depending on habitat conditions, varies in size of all plant sections as well as the extent of pubescence. A. Muldashev (Bot. zhurn. l.c.) separated form *A. fruticulosa* var. *kunlunica* Muld. from foothills of Kuen-Lun and Altyntag mountain range. This form differs only in the small size of the plant as a whole. He also described (Bot. zhurn. l.c.) one more variety *A. fruticulosa* var. *ramosissima* Muld. from Ordos which is characterised by loose corymbose inflorescence, a form well-preserved throughout the distribution range. Only in the foothills of Trans-Altay Gobi have we found bright green specimens of *A. fruticulosa* devoid of pubescence forming highly lignifying shoots and fairly compact mat with pinnatisected leaves and fairly loose corymbose inflorescence. In this respect, these specimens differ quite distinctly from type species. This is probably yet one more variety which we treat as *A. fruticulosa* var. *gobica* Filat.; its distribution range and ecological affinity are also well perceived.

5. **A. gracilis** (Hook. f. et Thoms.) Poljak. ex Tzvel. in Fl. URSS, 26 (1961) 407; Ikonn. Opred. rast. Pamira [Key to Plants of Pamir] (1963) 239; Opred. rast. Sr. Azii [Key to Plants of Mid. Asia] 10 (1993) 591. – *Tanacetum gracile* Hook. f. et Thoms. Fl. Brit. Ind. 3 (1881) 318.

Described from West. Himalayas. Type in London (K). Plate IV, fig. 2.

On rocky and stony slopes of high mountains and in rock crevices, 2700–3300 m alt.

IIIC. Pamir (Sarykol mountain range, Pistan gorge, Tadumbash-Pamir, July 15, 1901 – Alekseenko; Kok-Muinak pass, at exit from Tagarma valley, July 27, 1909 – Divn.; Burumsal area, 3000 m, Aug. 13; Pas-Rabat area, 3300 m, Aug.-Sept. 1941; Pakhtu river gorge, 2700 m, Aug. 2, 1942 – Serp.).

General distribution: East. Pamir, Himalayas, China (South-West.).

6. **A. grubovii** Muld. in Bot. zhurn. 11, 67 (1982) 1529; Gub. Konsp. Fl. Vneshn. Mong. [Conspectus of Flora of Outer Mongolia] (1996) 96.

Described from Mongolia (Jung. Gobi). Type in St.-Petersburg (LE). Plate IV, fig. 3. Map 3.

Petrophyte of rocky and stony slopes of low mountains; in petrophytic-forbs and feather grass-biurgun — *Ephedra* associations, rarely in river valleys, rock crevices.

IA. Mongolia: *Mong. Alt.* (pass in Dzhirgalantu river valley, Sept. 16, 1930 — Bar., paratypus!; Bulugun somon, Ded-Nariin-gol creek valley, rocky slope, July 27, 1947 — Yun.; Bulgan-gol river basin, Ded-Nariin-sala, 4 km from estuary, left bank of Bulgan-gol, 1550–1600 m, Aug. 17; Bulgan-gol gorge, 3–4 km beyond Ded-Nariin-sala estuary, 1540 m, left flank of gorge exposed northward, on granitic rocks, Aug. 17–1979, Grub., Muld., Dar., paratypus!).

IIA. Junggar: *Jung. Gobi* (Khara-Togo-ula, 15 km north of Altay somon, feather grass-biurgun association, Aug. 2; 15 km west of Uench somon, Dzagyn — ula, petrophyte — forbs association, Aug. 9–1977, Volk. et Rachk., paratypus!; Gobi dzhungarica, vallis fl. Uenczin gol, 4 km infra ostium Fl. Putzacty-gol, in declivi detritico ex saxis schistosis lateralis dextre desertum fruticosum in detritis gypsacio, Aug. 15, 1979, No. 1437 — Grubov, Muldaschev et al, typus!).

General distribution: endemic.

7. A. khantensis (Dunn) Shih in Acta Phitotax. Sin. 17, 2 (1979) 115; Fl. Sin. 76, 1 (1983) 113; Fl. Xizang. 4 (1985) 737; Fl. Intramong., ed. 2, 4 (1993) 591. — *Tanacetum khantense* Dunn, in Kew Bull. (1922) 150. — Ic.: Fl. Sin. l.c. tab. 17, fig. 3; Fl. Intramong. l.c. tab. 233, fig. 5–8.

Described from Qinghai. Type in London (K).

On rocky slopes, in arid beds of mountain rivers.

IA. Mongolia: *Alash. Gobi* ("spur of Khelan'shan' mountain range, Nin'sya" — Fl. Intramong. l.c.), *Ordos* (Fl. Intramong. l.c.), *Khesi* (Beishan' mountain range, arid steppe, Sept. 6; same site, Ludzhan-Dzhin town, Sept. 13–1890, Gr.-Grzh.; 100 km south-west of Dun'khuan, arid river bed in high foothills of Altyntag mountain range, Aug. 2, 1958 — Petr.).

General distribution: endemic in Central Asia.

8. A. nematoloba (Hand.-Mazz.) Ling et Shih in Bull. Bot. Lab. North.-East. Forest. Inst. 6 (1980) 16; Fl. Sin. 76, 1 (1983) 124; [Drevesn. rast. Tsinkhaya — Wooded Plants of Qinghai] (1987) 630; Fl. Intramong. ed. 2, 4 (1993) 590. — *Chrysanthemum nematolobum* Hand.-Mazz. in Acta Horti Gothob. 12 (1938) 271. — Ic.: Fl. Sin. l.c. tab. 16, fig. 2; Fl. Intramong. l.c. tab. 235, fig. 1–2; [Drevesn. rast. Tsinkhaya — Wooded Plants of Qinghai] l.c. tab. 449.

Described from Qinghai. Type in Beijing (PE). Map 3.

On steppe rocky slopes of high mountains.

IA. Mongolia: *West. Gobi* (Beishan' mountain range, Ludzhan-Dzhin town, steppe, Sept. 6, 1890 — Gr.-Grzh.); *Alash-Gobi* (Fl. Intramong. l.c.).

IC. Qaidam: *Mount.* (15 km south of Aksai settlement, Altyntag mountain range, rocky slopes of gorge, 2800 m alt., Aug. 2, 1958 — Petr.).

General distribution: China (South-West.).

9. A. parviflora (Gruning) Ling in Bull. Bot. Lab. North-East. Forest. Inst. 6 (1980) 15; Fl. Sin. 76, 1 (1983) 120; Fl. Intramong. ed. 2, 4 (1993)

589. — *Chrysanthemum parviflorum* Gruning in Feddes Repert. 12 (1913) 312; Hand.-Mazz. in Acta Horti Gothob. 12 (1938) 268; S.J. Hu in Quart. Journ. Taiwan Mus. 19 (1966) 39. — *Tanacetum parviflorum* (Gruning) Kung in Contr. Inst. Bot. Nat. Ac. Peiping, 2 (1934) 404. — *T. trifidum* auct. non DC.: Franch. in Nouv. Arch. Mus. Hist. natur. Paris, 6 (1883) 51 [Pl. David. 1 (1884) 172]. — Ic.: Fl. Intramong. l.c. tab. 233, fig. 1–4.

Described from Nor.-West. China. Type in Beijing (PE).

On rocky mountain slopes, on arid beds of mountain rivers.

IA. Mongolia: *Alash. Gobi* ("Alashan mountain range" — Fl. Intramong. l.c.); *Ordos* ("Ordos" Fl. Intramong. l.c.), *Khesi* (100 km south-west of Dunkhuan, arid river bed, Aug. 2; same site, on slopes of high foothills, Aug. 2–1958, Petr.).

IIIA. Qinghai: *Nanshan* ("vicinity of Nanashan" — Fl. Sin. l.c.).

General distribution: China (Nor.-West.).

10. A. przewalskii Poljak. in Not. syst. (Leningrad), 17 (1955) 422; Fl. Sin. 76, 1 (1983) 109; Fl. Intramong. ed. 2, 4 (1993) 591. — Ic.: Poljak. l.c. fig. 3; Fl. Intramong. l.c. tab. 235, fig. 5.

Described from Mongolia (Alash. Gobi). Type in St.-Petersburg (LE).

On rocky slopes of low mountains.

IA. Mongolia: *Alash. Gobi* (Alashan mountains, Aug. 9, 1880 — Przew., typus!; south. Alashan, near Dadzhin town, Aug. 9, No. 835 — Przew., paratypus!; "Helan'shan'" — Fl. Intramong. l.c.).

General distribution: China (North-West., South-West.).

11. A. purpurea Shih in Acta Phytotax. Sin. 17, 2 (1979) 115; Fl. Sin. 76, 1 (1983) 115; Fl. Xizang. 4 (1985) 738.

Described from Tibet (Lhasa). Type in Beijing (PE).

On mountain slopes, talus conglomerate and cliffs, rarely on cultivated lands.

IIIB. Tibet: *Weitzan* (Yangtze river basin, Nru-chyu area, on right bank of Golubaya river, on slopes and talus, 3500 m, July 25, 1900–Lad.), *South.* ("Nan'-Mulin, Lhasa" — Fl. Xizang. l.c.).

General distribution: China (South-West.).

12. A. roborowskii Muld. in Bot. zhurn. 67, 11 (1982) 528. — *Chrysanthemum stenolobum* Hand.-Mazz. in Acta Horti Gotob. 12 (1938) 265, nom. illegit.; Hu in Quart. Journ. Taiwan Mus. 19, 1–2 (1966) 43; Kitam. in Acta Phytotax Geobot. 23, 3–4 (1964) 74. — *Ch. tenuifolium* auct. non (Jacq.) B. Fedtsch.: Limpr. in Feddes Repert. 12 (1922) 507; Hu, l.c. (1966) 44. — *Tanacetum tenuifolium* auct. non Jacq.: Ling in Contr. Inst. Bot. Nat. Ac. Peiping, 2, 10 (1934) 502. — *Ajania tenuifolia* auct. non (Jacq.) Tzvel.: Iconogr. Cormophyt. Sin. 4 (1975) 516, icon. No. 6445.

Described from Qinghai (Qaidam). Type in St.-Petersburg (LE). Plate IV, fig. 1. Map 2.

In shrubby and cereal grass-forbs meadows on mountains.

IC. Qaidam: *Mount.* (Qaidam, "idoleum Dulan-Chit, 3400 m s.m. in argillosis et humosis, picecta et junipireta, Fl. flavi, Aug. 8, 1901, No. 394 — Lad.", typus!; alpine belt of Tetung mountain range,? 10–24, 1872, No. 290 — Przew., paratypus!).

IIIA. Qinghai: *Nanshan* (Rako-gol river, about 3000 m, July 7; between Nanshan and Rako-gol river, July 7–1880, Przew.; Nanshan, Dere-Chao, nor. bank of Kuku-nor lake, Aug. 4; same site, upper course of Khei-kho river, Aug. 22–1890, Gr.-Grzh.; Kuku-nor lake, Uiyu area, Aug. 13, 1908 — Czet., paratypus!; Mon' Yuan', Tatungkhe river valley (near stud farm), north-east. slope of conical hillock, cereal grass-forbs steppe, 2800 m, Aug. 20, 1958 — Dolgushin; Kuku-nor lake, meadow on east. bank, 3210 m, Aug. 5; 86 km west of Xining, pass, cereal grass-forbs steppe with alpine elements, Aug. 5; 25 km south of Lanzhou town, east. extremity of Nanshan, gently inclined slopes, 2435 m, mountain steppe with shrubs, Aug. 12–1959, Petr.).

IIIB. Tibet: Weitzan (on Bychu river — right tributary of Bychu, July 1–13, 1884 — Przew.; north-west. bank of Russkoe lake, on dry clay and in rock crevices, June 29, 1900 — Lad.).

General distribution: China (North-West., South-West.).

Note. Variety *A. roborowskii* var. *tsinghaica* Muld. described by A. Muldashev is distinguished by absence of rhizome and small size of plant as a whole. It has been reported from the following regions: Kuen-Lun, jugum Kukunoricum australe, fl. Baingol, 3600 m s.m. ad flumen, in glareosis, fl. flavi, July 26, 1894, No. 384 — Rob., typus!; regio Tangut, July 6, 1880, No. 489 — Prz., paratypus!; nor. bank of Kuku-nor lake, Degechu, Aug. 4, 1890, No. 229, paratypus!; Fl. Tetung, supra Junan-czen, ripa Aug. 8, 1890, No. 221 — Gr.-Grzh.; Tibet borealis, ad fl. dextrum fl. Botschu, 4200 m s.m., in graveosis ad rivules, parce, July 10, 1872 — Prz., paratypus!; Nanshan, upper course of Kheikho river, Aug. 22, 1890 — Gr.-Grzh.).

47

13. **A. scharnhorstii** (Rgl. et Schmalh.) Tzvel. in Fl. URSS, 26 (1961) 409; Fl. Kirgiz. 11 (1965) 150; Fl. Sin. 76, 1 (1983) 116; Fl. Xizang. 4 (1985) 738; [Drevesn. rast. Tsinkhaya-Wooded Plants of Qinghai] (1987) 630. — *Tanacetum scharnhorstii* Rgl. et Schmalh. in Acta Horti Petrop. 5, 2 (1878) 620. — *Hippolytia scharnhorstii* (Rgl. et Schmalh.) Poljak. in Not. syst. (Leningrad), 18 (1957) 289. — *Cryanthemum scharnhorstii* (Rgl. et Schmalh.) R. Kam. in Opred. rast. Sr. Azii [Key to Plants of Mid. Asia] 10 (1993) 638, 602. — Ic.: Tzvel. l.c. Plate 18, fig. 1; Fl. Sin. l.c. tab. 18, fig. 2; [Drevesn. rast. Tsinkhaya — Wooded Plants of Qinghai] l.c. Plate 45.

Described from Cent. Tien Shan. Type in St.-Petersburg (LE). Map 3.

On rubbly and rocky slopes, pebble beds in river valleys, 3000 m alt.

IIA. Junggar: *Tien Shan* (Bogdo-ula mountain, near Urumchi, Aug. 26–29, 1908 — Merzb.).

IIIB. Tibet: *Chang Tang* ("Zhitu, Getszi" — Fl. Xizang. l.c.).

General distribution: Cent. Tien Shan; endemic in Cent. Asia.

14. **A. tibetica** (Hook. f. et Thoms.) Tzvel. in Fl. URSS, 26 (1961) 410; Ikonn. Opred. rast. Pamira [Key to Plants of Pamir] (1963) 236; Fl. Sin. 76, 1 (1983) 115; Fl. Xizang. 4 (1985) 737. — *Tanacetum tibeticum* Hook. f. et Thoms. in Clarke, Comp. Ind. (1876) 154; Hook. f. Fl. Brit. Ind. 3 (1881)

391. — *Chrysanthemum tibeticum* (Hook. f. et Thoms.) Hoffm. in Pauls. Pl. coll. in Asia Med. and Pers. (1903) 149. — *Cryanthemum tibeticum* (Hook. f. et Thoms.) R. Kam. in Opred. rast. Sr. Azii [Key to Plants of Mid. Asia] 10 (1993) 639, 602. — Ic.: Ikonn. l.c. Plate 29, fig. 3.

Described from West. Himalayas (Nubra). Type in London (K).

On rocky and rubbly alpine slopes (3500–5200 m), rarely on arid beds of rivers.

IA. Mongolia: *Khesi* (nor. foothills of Nanshan, July 1879 — Przew.).

IC. Qaidam: *Mount.* (Ichegyn-gol river, on rocky arid bed of river, June 20, 1895 — Rob.).

IIIA. Qinghai: *Nanshan* (Humboldt mountain range, June 26, 1894 — Rob.).

IIIB. Tibet: *Chang Tang* (Russky mountain range, south. slope, vicinity of Dash-Kul' lake, on rocky slopes, June 30, 1889 — Rob.; Keriya (Russky mountain range), about 4000 m, July 13, 1885 — Przew.), *Weitzan* (nor. slope of Burkhan-Budda mountain range along Khatu river, July 31, 1884 — Przew.).

IIIC. Pamir (Kok-Muinak pass, on clayey-rocky slopes, July 8; Uicheklik river valley, on rubbly slopes, July 28-1909, Divn.; Tynnen-Davan pass, 4000–4200 m, June 26; Kara-Dzhilga river, alpine meadow, 4000–4500 m, July 22; Tespetlyk pass, about 5000 m, July 25; Goo-Dzhiro river, 4500–5500 m, July 27; same site, Shorluk river, 4000–5500 m, July 27; same site, Issyk-su river, on riverine pebble beds, Aug. 3; upper Kanlyk river, alpine tundra, Aug. 14; in Yazlek river valley (right tributary of Shinda river), Aug.-Sept. 1942, Serp.).

General distribution: Himalayas, Hindukush.

15. A. trifida (Turcz.) Tzvel. in Fl. URSS, 26 (1961) 412; Grub. Opred. rast. Mong. [Key to Plants of Mongolia] (1982) 245; Muld. in Bot. zhurn. 68, 2 (1983) 213; Gub. Konsp. Fl. Vneshn. Mong. [Conspectus of Flora of Outer Mongolia] (1996) 96. — *Artemisia trifida* Turcz. in Bull. Soc. natur. Moscou, 5 (1832) 196. — *Hippolytia trifida* (Turcz.) Poljak. in Not. syst. (Leningrad), 18 (1957) 289; Fl. Sin. 76, 1 (1983) 88.

Described from Mongolia (East. Mong.). Type in Kiev (KW). Isotype in St.-Petersburg (LE). Map 3.

Desert-steppe species distributed in feather grass-alliaceous, feather grass — tansy associations on rocky and stony slopes of low mountains and conical hillocks, hummocky sand, rarely on solonetzic soils.

IA. Mongolia: *East. Mong.* ("In montosis lapidosis Mongolia chinensis", 1831 (fl.). — I. Kuznetzov, isotypus!; vicinity of Dariganga, Tukhumyin-Gobi, Aug. 27; same site, vicinity of Ikhe-bulak, Uizichiin-Gobi, Aug. 23-1927, Zam.; Dariganga, 25 km north-east of Argaleul mountain, in clay basin, Sept. 1, 1931 — Pob.), *Val. Lakes* (3 km north of Khank urton, feather grass — tansy steppified desert, on loamy sand, Sept. 20, 1943 — Yun.; 49 km north of Khovd somon, intermontane plain, July 27, 1970 — Karam., Lavr., Bonzragch et al), *Gobi Alt.* (Artsa-Bogdo mountain range, near Kotel'-usu-khuduk collective, Aug. 7, 1926 — Glag.; Baga-Bogdo mountains, Sept. 10, 1949 — Eregdyn-Dagva), *East. Gobi* (Udinskaya lowland before Ubugun-Tsagan-obo, Aug. 26, 1928 — Shastin; Kalgan road between Sain-usu well and Toskho-nur lake, Aug. 15; same site, 17 km north of Dzamyin-ude, on Khukh-Tologoi mountain slopes, Aug. 28 — 1931, Pob.;

south-west. extremity of Sain-usu area, tansy-feather grass desert steppe on solonetzic soils, Aug. 28; 16 km north-east of Sain-Shanda, basin of Tel'-ulan-Shanda lake, fine hummocky sand, Aug. 31–1940, Yun.; 60 km east of Ulan-Badarkhu somon, on old caravan road, along slopes of rocky conical hillocks covered with sand, 1941; 5–6 km west of Agaratu-khuduk, on road to Khatun-bulak somon, feather grass-tansy desert steppe, June 30; 30 km north-east of Undur-Shili somon Deng-khuduk well, feather grass–tansy desert steppe, July 4–1941, Yun.; 3 km north of Dulan-Ikhe, Shandy-khuduk, alliaceous–feather grass steppe, Sept. 19. 1949–Yun.; 50 km east–north-east of Sain-Shanda, July 20, 1971–Kerzhner; 53 km south–south-east of railway siding No. 25, on top of conical hillock, July 15; 30 km south–south-east of Khubsugul, feather grass steppe, July 27–1971, Isach. et Rachk.; southward of Khotan-Bulak, nor. part of Khugde-ul, rocky slopes and rocks, Aug. 4, 1989–Grub., Gub. et Dar.), *Alash.-Gobi* (conical hillocks at border 20 km south of Shuulin post, nor. rocky slopes, Aug. 1, 1989–Grub., Gub. et Dar.), *Ordos* (Fl. Intramong. l.c.).

General distribution: endemic.

16. A. trilobata Poljak. ex Tzvel. in Fl. URSS, 26 (1961) 880, 409; Fl. Kirgiz. 11 (1965) 151; Fl. Sin. 76, 1 (1983) 116.— *Cryanthemum trilobatum* (Poljak. ex Tzvel.) R. Kam. in Opred. rast. Sr. Azii [Key to Plants of Mid. Asia] 10 (1993) 636, 602.

Described from Cent. Tien Shan (Sary-Dzhas river basin). Type in St.-Petersburg (LE).

Species of desert steppes made up of feather grass-tansy associations on rocky slopes of mountains, up to 2500 m alt., rarely on pebble beds of arid water channels.

IA. Mongolia: *Gobi Alt.* (4–5 km north of Khong urton, vicinity of Khara-Toirim lowland, feather grass-tansy desert steppe on solonetzic loamy sand, Sept. 20, 1943–Yun.); *East. Gobi* (south-west. extremity of Sain-usu area, desert steppe, July 28; 16 km north-east of Sain-Shanda, Tel'ulan-Shanda area, tansy-feather grass desert steppe, Aug. 31; 16 km north-east of Sain-Shanda on road to Baishintu, feather grass–tansy desert steppe, Sept. 2–1940, Yun.).

IIA. Junggar: *Jung. Gobi* ("Junggar basin"–Fl. Sin. l.c.).

General distribution: Cent. Tien Shan; endemic in Cent. Asia.

16. Hippolytia Poljak.
in Not. syst. (Leningrad), 18 (1957) 288, p.p.
— *Tanacetum* auct. non L., p.p.

1. Anthodia form simple or compound corymb at tip of stem; phyllary herbaceous, outer phyllary broad-lanceolate .... 2.

+ Anthodia in 2–5 numbers form subcapitate corymb at tip of stem; phyllary coriaceous-herbaceous, outer phyllary narrow-ovoid .... 3.

2. Fertile shoots somewhat leafy; lower cauline leaves on petioles longer than leaf blade, enlarged at base; phyllary with dark-brown margin .... 2. H. herderi (Rgl. et Schmalh.) Poljak.

+   Fertile shoots compactly leafy; lower cauline leaves on short petioles, not exceeding leaf blade, not enlarged at base; phyllary with light brownish margin .... 1. H. desmantha Shih.

3.   Plants with thick woody root forming many fertile and vegetative shoots; leaf blade of lower cauline leaves oblong-ovoid in profile, simple-pinnatisected .... 4.

+   Plants with multicipital, somewhat lignifying rhizome forming rather few (3–5) fertile shoots; leaf blade of lower cauline leaves broad-obovoid in profile, 2-pinnatisected .... 5.

4.   Anthodia on pedicels up to 4 cm long at tip of fertile shoots; phyllary somewhat pubescent, with sparse tomentose pubescence only at base, outer phyllary broad-lanceolate-ovoid .... 4. H. senecionis (Jacq. ex Bess.) Poljak.

+   Anthodia subsessile, form compact subcapitate inflorescence at tips of fertile shoots; phyllary with compact white tomentose pubescence, outer phyllary oblong-ovoid ....
.... 5. H. sincalathiiformis Shih.

5.   Fertile shoots branched in upper portion; blades of lower cauline leaves greyish due to compact sparse tomentose pubescence; terminal leaf lobules subobtuse; anthodia form compact, compound corymb at tip of fertile shoots ....
.... 3. H. glomerata Shih.

+   Fertile shoots not branched, terminate at tip in compact, subcapitate inflorescence; blade of lower cauline leaves greyish green, with fine tomentose pubescence; terminal leaf lobules subacute .... 6. H. tomentosa (DC.) Tzvel.

1. H. desmantha Shih in Acta Phytotax. Sin. 17, 4 (1979) 63; Fl. Sin. 76, 1 (1983) 90; [Drevesn. rast. Tsinkhaya — Wooded Plants of Qinghai] (1987) 627. —Ic.: Shih, l.c. tab. 1, fig. 3.

Described from Nor. Tibet (Qinghai province, Yuishu). Type in Ugyan (NWBI).

On arid loessial slopes, along fringe of cliffs, in rock crevices, 3500–4000 m alt.

IA. Mongolia: *Khesi* (20 km north-west of Lanzhou town on road to Yuimyn', steppe along loessial slope, Oct. 9, 1957 — Yun.).

IIIB. Tibet: *Weitzan* ("Yuishu" — [Drevesn. rast. Tsinkhaya — Wooded Plants of Qinghai] l.c.).

General distribution: endemic in Cent. Asia.

2. H. glomerata Shih in Acta Phytotax. Sin. 17, 4 (1979) 67; Fl. Sin. 76, 1 (1983) 94; Fl. Xizang. 4 (1985) 732.

Described from South. Tibet (Nimu). Type in Beijing (PE).

On rocky slopes, rocks, 3500 m and above.

IIIB. Tibet: *South.* ("Nimu" — Fl. Sin. and Fl. Xizang. l.c.).

General distribution: endemic in Tibet.

3. H. herderi (Rgl. et Schmalh.) Poljak. in Not. syst. (Leningrad), 18 (1957) 289; Tzvel. in Fl. SSSR, 26 (1961) 414; Fl. Kirgiz. 9 (1965) 152; Fl. Kazakhst. 9 (1966) 74; Fl. Sin. 76, 1 (1983) 92; Opred. rast. Sr. Azii [Key to Plants of Mid. Asia] 10 (1993) 600. — *H. leucophylla* (Rgl.) Poljak. l.c. 289. — *Tanacetum herderi* Rgl. et Schmalh. in Acta Horti Petrop. 5, 2 (1878) 619. — *T. leucophyllum* Rgl. in Acta Horti Petrop. 7, 2 (1881) 551. — Ic.: Fl. Kazakhst. l.c. Plate 9, fig. 4.

Described from East. Kazakhstan (Trans-Ili Ala Tau). Type in St.-Petersburg (LE). Plate IV, fig. 5. Map 3.

On rocky and stony slopes, 2500 m alt. and above.

IIA. Junggar: *Tien Shan* (Bogdo mountain, Kok-Kamyr, about 3000 m, July 25, 1878; Kash, Monguto area, July 4; Kunges, Aug. 27, 1879 — A. Reg.; Mukhurdai river, 3500–3700 m, July 18, 1893 — Rob.).

General distribution: Jung.-Tarb., Nor. and Cent. Tien Shan, Mid. Asia (West. Tien Shan, Pam.-Alay).

4. H. senecionis (Jacq. ex Bess.) Poljak. in Fl. URSS, 26 (1961) 414, in nota; Fl. Sin. 76, 1 (1983) 93; Fl. Xizang. 4 (1985) 731. — *Artemisia senecionis* Jacq. ex Bess. in Bull. Soc. natur. Moscou, 9 (1836) 75. — *Tanacetum senecionis* (Jacq. ex Bess.) J. Gay in DC. Prodr. 6 (1838) 129.

Described from Himalayas. Type in Paris (P).

On rocky and stony mountain slopes, 2000–2800 m alt.

IIIB. Tibet: *Weitzan* (I-chyu river valley near Achokak-Gomba monastery (Goluboi river basin), July 27, 1900 — Lad.).

General distribution: South-West. China (Sikang).

5. H. sincalathiiformis Shih in Acta Phytotax. Sin. 17, 4 (1979) 66; Fl. Sin. 76, 1 (1983) 94; Fl. Xizang. 4 (1985) 733. — Ic.: Shih, l.c. tab. 1, fig. 1.

Described from South. Tibet (Nimu). Type in Beijing (PE).

On rubbly and rocky slopes of high mountains, 4500 — 5500 m alt.

IIIB. Tibet: *South.* ("Lhasa, Nimu Tszyacha" — Fl. Sin. (1983) and F. Xizang. (1985) l.c.).

General distribution: endemic in Cent. Asia.

Note. It is possible that one more species, *H. kennedyi* (Dunn) Ling, grows in the alpine southern Tibet, also in Tszyacha of Lhasa region. Although this species has been reported in Fl. Sinica and Fl. Xizangica, it is not being treated in this compendium for want of herbarium material. This species is extremely similar to *H. sincalathiiformis* in diagnostic features and ecology.

6. H. tomentosa (DC.) Tzvel. in Fl. URSS, 26 (1961) 416, in nota; Shih in Acta Phytotax. Sin. 17, 4 (1979) 66; Fl. Sin. 76, 1 (1983) 93; Fl. Xizang. 4 (1985) 731. — *Tanacetum tomentosum* DC. Prodr. 6 (1838) 130; Clarke, Comp. Ind. (1876) 155. — Ic.: Fl. Sin. l.c. tab. 14, fig. 2.

Described from Himalayas. Type in Geneva (G).

On rocky slopes, in rock crevices, 3500–3700 m alt.

IIIB. Tibet: *South.* (Ali region — Fl. Sin. l.c. and Fl. Xizang. l.c.)

General distribution: Himalayas (west.).

17. Filifolium Kitam.
in Acta Phytotax. et Geobot. 9 (1940) 157

1. F. sibiricum (L.) Kitam. in Acta Phytotax. Geobot. 9 (1940) 157; Tzvel. in Fl. SSSR, 26 (1961) 417; Grub. Opred. rast. Mong. [Key to Plants of Mongolia] (1982) 245; Fl. Sin. 76, 1 (1983) 128; Fl. Intramong. ed. 2, 4 (1993) 593; Gub. Konsp. fl. Vneshn. Mong. [Conspectus of Flora of Outer Mongolia] (1996) 102. — *Tanacetum sibiricum* L. Sp. pl. (1753) 844; DC. Prodr. 6 (1838) 129. — Ic.: Gmel. Fl. Sib. 2 (1749) 134, tab. 65, fig. 2; Tzvel. l.c. Plate 19.

Described from East. Siberia. The cited Gmelin's sketch represents the type.

Steppe species, distributed in cereal grass-forbs, cereal grass-tansy, and wormwood-tansy associations in plains and lower mountain belts.

IA. Mongolia: *Cent. Khalkha* (Muren somon, 20–25 km north of Underkhan, wormwood-cereal grass steppe, July 25, 1949 — Yun.; 60 km south of Underkhan, forbs-sheep's fescue steppe, July 9, 1971 — Isach. et Rachk.; 42 km west — south-west of Underkhan, rocky slope, July 18, 1971 — Dashnyam, Isach., Karam. et al), *East. Mong.* (Kulun-Buinur plain, Elesyn-khuduk collective, clayey-sandy soil, July 9; same site, Khaligakha area — Abder river, July 23; same site, between Kyrymty and Buin-gol, Aug. 18–1899, Pot. et Sold.; Khailar town, July 6, 1901 — Lipsk.; vicinity of Manchuria station, June 6, 1902 — Litw.; Manchuria station, 1915 — Nechaeva; same site, high arid site on sandy plain, June 20, 1951 — A.R. Lee (1959); vicinity of Manchuria station, arid site on mountain slope, June 24, 1951 — Wang Chang; Khukh-Khoto, 10 km from town, northern foothills of Datsin'-Shan', feather grass-wild rye steppe, June 4, 1958 — Petr.; Sara-Tol State Farm, 40 km north-west of Khailar town, wormwood-forbs steppe, Sept. 11, 1958 — Lavr. et al; Shilin-Khoto, steppe, 1959 — Ivan.; 67 km south of Bodonchin-khuduk, petrophyte-forbs association, June 30; 33 km north-west of Khutliin-khuduk, forbs-sheep's fescue steppe, July 14–1971, Dashnyam, Isach., Karam., Rachk. et Safronova).

General distribution: East. Sib., Far East, Nor. Mong., Chima (Dunbei), Korea.

18. Kaschgaria Poljak.
in Not. syst. (Leningrad), 18 (1957) 282

1. Fertile shoots with long declinate branchlets; leaf blade simple, lanceolate-filiform-linear; anthodia gathered into clusters at

ends of fertile shoots ....
.... 1. K. brachanthemoides (Winkl.) Poljak.

+ Fertile shoots with short branchlets; leaf blade tri- or pinnatisected; anthodia gathered into compact corymbose or subfascicular inflorescence ....
.... 2. K. komarovii (Krasch. et Rubtz.) Poljak.

1. K. brachanthemoides (Winkl.) Poljak. in Not. syst. (Leningrad), 18 (1957) 283 and in Fl. URSS, 26 (1961) 424; Fl. Kazakhst. 9 (1966) 75; Fl. Sin. 76, 1 (1983) 129; Opred. rast. Sr. Azii [Key to Plants of Mid. Asia] 10 (1993) 535. — *Artemisia brachanthemoides* Winkl. in Acta Horti Petrop. 9, 2 (1886) 422. — *Tanacetum brachanthemoides* (Winkl.) Krasch. in Acta Inst. Bot. Ac. Sci. URSS, 1, 1 (1933) 175; Krasch. in Not. syst. (Leningrad), 9 (1946) 168. — Ic.: Fl. SSSR, l.c. Plate 20, fig. 2; Fl. Kazakhst. l.c. Plate 9, fig. 5.

Described from Sinkiang (Tien Shan). Type in St.-Petersburg (LE). Map 4.

On rocks on barren low mountains, old arid river beds, 1000 m alt. and above.

IB. Kashgar: *East.* (Turfan town, Algoi river, Aug. 13, 1879 — A. Reg., syntypus!; Khami desert, Shugud well, along old Goly-Toga river beds, Aug. 11, 1895 — Rob.; Chokur area, Aug. 28; Choltag mountains, Argyi-bulak gorge, Aug. 31–1929, Pop.).

IIA. Junggar: *Tien Shan* (Borborogusun, 1000 m alt. and above, Aug. 25, 1878 — A. Reg., lectotypus!).

General distribution: Nor. Tien Shan; endemic in Central Asia.

2. K. komarovii (Krasch. et Rubtz.) Poljak. in Not. syst. (Leningrad), 18 (1957) 283 and in Fl. URSS, 26 (1961) 424; Fl. Kazakhst. 9 (1966) 75; Grub. Opred. rast. Mong. [Key to Plants of Mongolia] (1982) 245; Fl. Sin. 76, 1 (1983) 129; Opred. rast. Sr. Azii [Key to Plants of Mid. Asia] 10 (1993) 536; Gub. Konsp. fl. Vneshn. Mong. [Conspectus of Flora of Outer Mongolia] (1996) 103. — *Tanacetum komarovii* Krasch. et Rubtz. in Not. syst. (Leningrad) 9 (1946) 168; Kryl. Fl. Zap. Sib. 11 (1949) 2759. — *Chrysanthemum komarovii* (Krasch. et Rubtz.) S.J. Hu in Quart. Journ. Taiwan Mus. 19 (1966) 30. — Ic.: Fl. Kazakhst. l.c. Plate 9, fig. 2.

Described from Mongolia (Jung. Gobi). Type in St.–Petersburg (LE). Map 4.

On rocky slopes of barren low mountains, rocky floors of gorges, talus, rocks, up to 2000 m alt.

IA. Mongolia: *Mong. Alt.* (Bulgan somon, Ulyastyin-gol river, 20 km north of somon on rocks, July 1, 1980 — Kerzhner; 25 km north of Bulgan settlement, rocky slope toward Ulyastyiin-gol river (left tributary of Bulgan river), 7 km from estuary, Aug. 29, 1988 — Kam., Gub., Dar.), *West. Gobi* (Bain-Gobi somon, foothills of Tsagan-Bogdo mountain range, along rocky hummocky area, Aug. 1, 1943 — Yun.; Atas-ula, on rocks, July 24;

Bulgan-Khoshuin-nuru mountain range, in deep gorge, July 26, 1973 — Isach. et Rachk.; Ara-Khure-Bayan-nuru, sandy-rocky slope, Aug. 5, 1973 — Golubk. et Tsogt; Tsagan-Bogdo, nor. slope at 1950 m alt., *Sympegma*-feather grass association, July 19; Ederingiin-nuru mountain range, in small ravine, July 27; 80 km south — south-west of Dzakhoi oasis, on slopes of conical hillocks, Aug. 16–1973, Isach. et Rachk.; south-west. fringe of Chingiz-ul, rocky floor of gorge, Aug. 19, 1973 — Rachk. et Damba; 3 km west of Dzakhoi oasis, granitic Khatan-Khairkhan-ul massif, rocky floor of gorge, July 25, 1977 — Volk. et Rachk.; 80 km south — south-west of Bayan-Tsagan and Khairkhan-Sairiin-ula, biurgun-*Sympegma* association, Aug. 6, 1979 — Isach. et Rachk.; solonchak ridge south of Ergiin-us spring on border road, shrubby formation along rocky floor of gorge, Aug. 26; Tsagan-Bogdo mountain range, south. macroscope, 13 km on road to Ekhin-gol from Tsagan-bulak spring, on sandy-rocky gorge, Aug. 28–1979, Grub., Muld. et Dar.).

IIA. Junggar: *Tarb.* (Kotbukha area, Aug. 10, 1876–Pot.); *Tien Shan* (Kuruk-Tag, between Turfan and Karashir basin, pass through mountain range in the form of flat saddle-shaped valley, rocky barren land, along flanks of gorges, Sept. 16, 1957 — Yun., Li et Yuan'); *Jung. Gobi* (barren steppe between Yamatei and Baityk-Bogdo mountains, Aug. 6, 1898 — Klem., typus!; south. Altay-Pevtsov, paratypus!; Saepu lake and settlement, from Urumchi in Turfan oasis, Sept. 5, 1929 — Pop.; Bulugun somon, Oshigiin-usu, granitic hummocky area, July 30; same site, Demchigiin-Khuren-Undur mountains, 20 km east of Oshigiin-usu area, gorge in saxaul desert, July 31–1947, Yun.; Baitag-Bogdo mountain range, left creek valley of Ulyastu-gol gorge 7 km from estuary along nor. slope at 2000 m alt., Sept. 17; south. foothills of Barangiin-Khara-nuru mountain range, 10 km west — south-west of Mergen-ula, along floor of gorge, Sept. 19–1948, Grub.; 83 km south of Ertai on road to Guchen, rocky slope, July 16, 1959 — Yun., Yuan'; Bulgan somon, Dzagin-Ulan low mountains, July 17 — Kam. et Dar.; nor. slope of Baitag-Bogdo mountain range, lower portion of Tsagan-Burgastyn-gol gorge, 25 km east of Baitag-Bogdo post, Aug. 4, 1988 — Kam., Gub. et Dar.).

General distribution: Jung. Tarb., Nor. Tien Shan; endemic in Cent. Asia.

19. Neopallasia Poljak.
in Not. syst. (Leningrad), 17 (1955) 429

11. N. pectinata (Pall.) Poljak. in Not. syst. (Leningrad) 17 (1955) 428 and in Fl. URSS, 26 (1961) 637; Fl. Sin. 76, 1 (1983) 130; Fl. Xizang. 4 (1985) 740; Fl. Intramong. ed. 2, 4 (1993) 667; Opred. rast. Sr. Azii [Key to Plants of Mid. Asia] 10 (1993) 588; Gub. Konsp. fl. Vneshn. Mong. [Conspectus of Flora of Outer Mongolia] (1996) 103. — *Artemisia pectinata* Pall. Reise, 3 (1776) 755; Kryl. Fl. Zap. Sib. 11 (1949) 2816; Fl. Kirgiz. 11 (1965) 164; Fl. Kazakhst. 9 (1966) 96; Grub. in Opred. rast Mong. [Key to Plants of Mongolia] (1982) 246. — Ic.: Fl. SSSR, l.c. Plate 31, fig. 1; Fl. Kazakhst. l.c. Plate 10, fig. 2; Fl. Intramong. l.c. tab. 229, fig. 1–3.

Described from East. Siberia (south. Trans-Baikal). Type in St.-Petersburg (LE).

In desert-steppe zone, on loamy, loamy-sandy plains, dried up lakes, rocky trails, talus, gorge floors.

IA. Mongolia: *Mong. Alt.* (steppe around Kobdo town, July 18, 1906 — Sap.; valley of Tsinkir-Tyuguryuk river, Aug. 13; Khara-Adzarga mountain range, valley of Urtu-Ulya sutai river, Aug. 27; Khasagtu-Khairkhan mountains, exit of Dundu-seren-gol river onto trail, Sept. 15; around Undur-khairkhan mountain (west. slope), Dundu-Seren-gol river, on rocky talus, Sept. 16–1930, Pob.; bank of Dzhirgalanta-gol river, arid bed, July 27, 1950 – Kuznetsov), *Cent. Khalkha* (vicinity of Ikhe-Tukhum-nor lake, bank, June 1926 – Zam.; Utat, Aug. 8, 1927 – Terekhovko; Alkhakhoshuin–Gobi, Aug.13; same site, dried up Bilkichen-nor lake, Aug. 14–1927, Zam. et Terekhovko; Erdeni-Dalai somon, old Ulan-Bator – Dalan-Dzadaged road 15 km north of somon, feather grass-alliaceous-snake weed steppe, July 13, 1948 – Grub.; Sumber somon, Choir state farm, 6 km north of Magakhyn-ul, desertified steppe, Aug. 11, 1970 – Mirkin, Kashanov et al), *East. Mong.* (Dariganga, vicinity of Tukhumyn-Gobi, Aug. 23 and Aug 26, 1927 – Zam.; same site, around Argaleul mountain, wormwood steppe, Aug. 31, 1931 – Pob.; 4 km north-west of Khongor, on trails, July 4, 1971 – Karam. et Safronova), *Depr. Lakes* (isthmus between Ubsu-nur lake and Baga-nur lake, Sept. 22, 1879 – Pot.; nor. bank of Ubsu-nur lake, desert steppe, July 3, 1892 – Kryl.; near Kobdo town, valley of Buyantu river, on weathered shales, Aug. 11 and Sept. 1, 1930 – Bar.; Shargiin-Gobi desert, from Taishir mountain range to Khalyun river, Aug. 16; Buyantu river, in plantations, Aug. 28–1930, Pob.; Borigingol area, along south. fringe of sand 4–5 km east of Dzun-Gobi somon, along margin of solonchak, July 26, 1945 – Yun.; east. trail of Umne-Khairkhan trail, desert steppe, July 26, 1972 – Metel'tseva; Tes river floodplain, 60 km east – north-east of Tes somon, in shrubs, June 17, 1978 – Karam., Beket. et al; 30 km south-east of south. bank of Khara-Us-nor, south-west.-west. slope of Dzhergalant-ula, rocky desert, July 24, 1979 – Gub.), *Gobi-Alt.* (Dundu-Saikhan mountains, July 13, 1909 – Czet.; Legin-gol river, sandy steppe, Aug. 1922 – Pisarev; Ikhe-Bogdo river, in lowlands, Aug. 18, 1926 – Tug.; Bain-Tsagan mountain, on arid bed, Aug. 5; Dundu-Saikhan mountain, on rubble on arid bed at exit of Ulan-Khundei creek valley, Aug. 18–1931, Ik.-Gal.; 2–3 km south of Noyan somon, desert steppe, July 5; piedmontane plain south of Ikhe-Bogdo mountain range, 10–15 km from Bayan-Gobi somon, desert steppe, Sept. 4–1943, Yun.; south. slope of Artsa-Bogdo mountain range, Ikhe-Bug-ula vicinity, on rubble slopes of mounds, July 21; south. slope of Ikhe-Bogdo mountain range, feather grass-alliaceous steppe, July 30; Ikhe-Bain-ula mountain range, in ravine with derris, Aug. 2; south. slope of Tostu-nuru mountain range, along pebble bed in ravine, Aug. 18; south. gently inclined trail of Gobi Altay 20 km north-west of Dzadagai-Khere, feather grass-biurgun steppe, Aug. 25–1948, Grub.; submontane plain on south. slope of Gurban-Saikhan, in plantations, Sept. 15, 1951 – Yun.; 40 km north of Obot settlement, Khuryn-Khoi-nuru mountains, July 25, 1972 – Rachk. et Guricheva; Gurban-Tes somon, 1 km north-east of Ekhin-Tsagan-Deris collective, sandy floor of gorge, Sept. 5, 1979 – Grub., Muld. et Dar.), *East. Gobi* (Argaliin mountain range, vicinity of Khodatyin-khuduk well, Aug. 5; northward of Yaman-Ikhe-Dulan-Khoshun, Aug. 20; Ton-Dzukha area, at exit of Tukhumyin-Gobi valley, Sept. 1–1928, Shastin; Khan-Bogdo somon, Khoir Ul'dzeitu area, near Khutag-ula, desert steppe, Aug. 23, 1930 – Kuznetsov; Del'ger-Khangoi mountains, steppe, July 30, 1931 – Ik. – Gal.; 75 km north-east of Dalan-Dzadagad on road to Ulan-Bator, desert steppe, Aug. 16; 40 km south of Dalan-Dzhirgalan somon, Delgerlin-Deris area, on patches of arid solonchak, Aug. 24; Delgeriin-Deris areas, 40 km south-east of Dalan-Dzhirgalan somon, desert steppe, Aug. 24; Durbul'dzhin-ula area, 47 km south-east of Choiroen on road to Sain-Shandu, desert steppe, Aug. 26, 6–7 km south – south-west of Sainusu collective, desert steppe, Aug. 28; 16 km north-east of Sain-Shanda, floor of solonchak lowland, Aug. 31; near Delgers somon, conical hillock slopes, Sept. 11; Abdaryntiin-Tsab basin, south of Sain-Shanda somon, floor of basin, Sept. 14; 30 km west of Sain-Shanda, floor of solonchak lowland, Sept. 17–1940, Yun.; Lus somon, Khatu-Tugrik area, lake terrace, July 12, 1950 – Lavr.; 10 km north of Sain-Shanda, plain, Aug. 26, 1950 – Ivan.; 10 km south-west of Bayan-Delger on road to Bayan-Munkh, wormwood-snakeweed association, July 6; 18 km north-east of Mandakh,

54

21. Elachanthemum Ling et Y.R. Ling
in Acta Phytotax. Sin. 16, 1 (1978) 62

1. E. intricatum (Franch.) Ling et Y.R. Ling in Acta Phytotax. Sin. 16, 1 (1978) 63; Fl. Sin. 76 1 (1983) 97; Fl. Intramong. ed. 2, 4 (1993) 581. — *Artemisia intricata* Franch. Pl. David. 1 (1884) 170; Grub. Opred. rast. Mong [Key to Plants of Mongolia] (1982) 265. — *Stilpnolepis intricata* (Franch.) Shih in Acta Phytotax. Sin 23, 6 (1985) 471; Gub. Konsp. fl. Vneshn. Mong. [Conspectus of Flora of Outer Mongolia] (1996) 107. — Ic.: Fl. Intramong. I.c. tab. 230, fig. 1–4.

Described from Inner Mongolia (Ulantsab). Type in Beijing (PE). Map 4.

In sandy and rubbly rocky deserts, along coastal pebble beds, rocky trails, gorges.

IA. Mongolia: *Mong. Alt.* (Khasagtu-Khairkhan town, rocky trail, Sept. 14, 1930 — Pob.), *East. Mong.* (35 km north-east of Argalsul mountain, in floodplain of Dzhingiin-bulak spring, Sept. 3; 25 km west of Baishiintu-Sume near Koshun-Shent collective, arid river valley, Sept. 9; vicinity of Baishiintu-Sume monastery, sandy bank of Tsagan-Irne lake, Sept. 10–1931, Pob.). *Depr. Lakes* (upper course of Botkhon river, coastal pebble beds, Sept. 8, 1930 — Pob.; sandy desert near weathering granites of Eshi mountain on descent from Khorgon to Dzergin-gol valley, Aug. 2, 1931 — Bar.; 50 km south of Kobdo, plain, alliaceous-biurgun desert with wormwood, Aug. 4, 1977 — Karam., Sanczir et Sumerina; north-east of Khara-Us-nur lake bank, rubbly-rocky desert, Sept. 3, 1978 — Gub.), *Val. Lakes* (south. bank of Orok-nur lake, Sept. 2; Tuin-gol river, Sept. 4–1886, Pot.; south, trail of Ikhe-Bogdo mountain range (midportion), desert steppe, Sept. 13, 1943 — Yun.), *Gobi Alt.* (Bogdo mountains, on sand, No. 306, 1925 — Chaney; foothills of Dzolin mountain range, rubbly steppe, Sept. 8; Dzolin mountains, on trail and in arid beds, Sept. 9; Bain-Tsagan mountains, on rubbly trail in Bain-Tukhum area, Sep. 11; Bain-Tukhum area, rubbly semidesert, Sept. 19–1931, Ik.-Gal.), *East. Gobi* (vicinity of Baga-Ude well, along slopes in Bain-gote mountain valley, Aug. 21; 17 km north of Dzamyn-Ude, on Khukh-Tologoi slope, Aug. 28–1931, Pob.; on rubbly trail of Del'ger-Khangai mountains, Sept. 26, 1931 — Ik.-Gal.; 32–35 km south-west of Ulan-khuduk collective on motorable road to Sain-Shande from Baishintu, saltwort-feather grass desert, Sept. 12, 1940 — Yun.; west. extremity of Delger-Khangai mountain range, 6 km south-west of Ulan-Khamar-khuduk well, granitic conical hillocks, Sept. 5, 1950 — Yun., Lavr., Kal.), *Alash. Gobi* (Sept. 1–2, 1880; Alashan mountains, Sept. 10, 1880 — Przew), *Khesi* (Yuimyn' town, Tsi-Lan'-Shan' mountains (west. part of Richthofen mountain range), steppified saltwort desert along foothills, Sept. 22, 1958 — Lavr. et al).

IIA. Junggar: *Jung. Gobi* (Ikhe-Alag-ula town, on road to Khairkhan somon from Altay somon, 43–47 km south-east of former and 12–16 km from gorge beginning, pebble bed — sandy floor of ravine, Aug. 20, 1979 — Grub., Muld., Dar.).

General distribution: Nor. Mong. (Hang., Mong.-Daur.), China (North-West.: Shenxi).

22. Artemisia L.

Sp. pl. (1753) 845 and
Gen. pl. ed. 5 (1754) 357

1. Anthodia heterogamous, with rather few marginal pistillate flowers and many bisexual or only staminate disc flowers ..... 2.

+ Anthodia homogamous with rather few (8–15) bisexual flowers without marginal pistillate flowers.. ....
.... Subgenus 3. Seriphidium (Bess.) Peterm.

2. Disc flowers invariably staminate but with reduced pistil..
.... Subgenus 2. Dracunculus (Bess.) Peterm.

+ All disc flowers bisexual .... Subgenus 1. Artemisia.

Subgenus 1. Artemisia

1. Perennial herbaceous plants and subshrubs .... 2.

+ Annual-biennial plants .... 58.

2. Blades of lower cauline leaves entire, simple or serratodentate or rather shallowly (not up to midrib) pinnatifid along margin into lobules of different length and breadth .... 3.

+ Blades of lower cauline leaves invariably pinnatisected up to midrib .... 23.

3. Blades of all leaves simple, entire, dentate or rather shallowly lobed .... 4.

+ Blades of lower and middle cauline leaves rather shallowly pinnati- or trisected .... 6.

4. All leaves linear, up to 1 cm broad, entire, wtih short linear auricles at base .... 55. A. subulata Nakai.

+ Lower and middle cauline leaves linear-lanceolate or lanceolate 1–2 cm broad, incised-dentate along margin, usually without auricles or with broad-lanceolate auricles at base .... 5.

5. Fertile shoots reddish violet, glabrous. Lower and middle cauline leaves lanceolate or linear-lanceolate, with rather small deltoid subacute teeth along margin. Anthodia nutant ....
.... 28. A. integrifolia L.

+ Fertile shoots green, pubescent, specially in lower portion. Lower and middle cauline leaves broad-lanceolate, with lanceolate or linear obtuse-acuminate teeth along margin. Anthodia not nutant .... 53. A. stolonifera (Maxim.) Kom.

6. Lobules of lower and middle cauline leaves narrow-linear or lanceolate, up to 2 mm broad, entire or narrow-dentate along margin .... 7.

+ Lobules of lower and middle cauline leaves relatively broad, 2–8 (10) mm broad, usually deeply lobately incised, rarely entire .... 8.

7. Fertile shoots 75–120 cm tall, somewhat pubescent, sub-glabrous. Middle and cauline leaves 8–12 cm long, terminal leaf lobules 4-6 cm long, tapered-acuminate .... 46. A rubripes Nakai.

+ Fertile shoots 50–80 cm tall, compactly pubescent with short appressed hairs. Middle cauline leaves 5–8 cm long; terminal leaf lobules 2–3 (4.5) cm long, short-acuminate .... 20. A. feddei Levl. et Vaniot.

8. Terminal leaf lobules with short acute teeth along margin. Anthodia campanulate, gathered into compact spicate inflorescences on lateral branchlets and form long narrow-pyramidal panicle .... 10.

+ Terminal leaf lobules without teeth along margin. Anthodia narrow-campanulate or terete, aggregated into loose fascicular inflorescences on lateral branchlets and form compact pyramidal panicle .... 9.

9. Fertile shoots glabrous, somewhat sulcate. Bracts not longer than anthodia. Outer phyllary orbicular or oval, somewhat cobwebby-pubescent or subglabrous .... 50. A. selengensis Turcz. ex Bess.

+ Fertile shoots pubescent with short appressed hairs, ribbed-sulcate. Bracts longer than anthodia. Outer phyllary elliptical, compactly cobwebby-pubescent, whitish .... 61 A. umbrosa (Bess.) Pamp.

10. Leaf blade coriaceous, glabrous on both sides. Panicle pyramidal, compactly leafy .... 60. A. tangutica Pamp.

+ Leaf blade soft, pubescent on both sides or only underneath. Panicle of different form, somewhat leafy .... 11.

11. Leaf blade glabrous on top, green or rarely pubescent with long appressed hairs, greyish or whitish on underside due to compact tomentose pubescence .... 12.

+ Leaf blade greyish on top due to fairly compact dense pubescence of short appressed hairs, with compact white tomentum underneath .... 20.

12. Rhizome thickened in upper portion. Fertile shoots 70–150 (200) cm tall. Leaf blade somewhat incised into broad-lanceolate lobules .... 13.

+ Rhizome not thickened in upper portion. Fertile shoots 40–70 (80) cm tall. Leaf blade deeply incised into narrow-linear lobules .... 16.

13. Lower cauline leaves petiolate, rest sessile. Anthodia 3–4 mm long; compact racemose inflorescences form panicles on lateral branchlets .... 14.

+ All leaves (excluding bracts) petiolate. Anthodia up to 2 (3) mm long, form panicles on lateral branchlets singly or 2–3 (5) together .... 15.

14. Fertile shoots brown-violet. Lower and middle leaves dark green above, subglabrous, greyish underneath due to cobwebby-tomentose pubescence. Anthodia oblong or narrow-campanulate .... 68. A. vulgaris L.

+ Fertile shoots light-brownish-green, somewhat pubescent. Lower and middle cauline leaves green on top with fine hairly pubescence, whitish cobwebby underneath. Anthodia broad-campanulate .... 27. A. igniaria Maxim.

15. Rhizome long, decumbent, with underground stunted shoots. Fertile shoots angular-ribbed. Lower and middle cauline leaves pinnatifid into linear subobtuse lobules. Anthodia sessile, upright .... 64. A. verlotiorum Lamotte.

+ Rhizome short, thickened in upper portion. Fertile shoots orbicular, somewhat sulcate. Lower and middle cauline leaves pinnatifid into lanceolate, denticulate lobules. Anthodia stalked, nutant .... 58. A. sylvatica Maxim.

16. Cauline leaves on petioles up to 2 cm long. Anthodia 3–4 mm long, single or 2–3 together in panicles on lateral branchlets .... 66. A. vexans Pamp.

+ Cauline leaves on 3–7 cm long petioles. Anthodia up to 3 mm long, gathered in bunches of compact racemose inflorescences and form panicles on lateral branchlets .... 17.

17. Fertile shoots angular-sulcate, dark brown. Terminal leaf lobules sharp-toothed along margin. Panicle narrow-pyramidal .... 18.

+ Fertile shoots somewhat ribbed, light brown. Terminal leaf lobules without teeth or blunt-toothed along margin. Panicle broad-pyramidal .... 40. A. obscura Pamp.

18. Fertile shoots glabrous or diffusely pubescent. Terminal leaf lobules oblong-lanceolate or narrow-lanceolate. Anthodia campanulate .... 38. A. mongolica (Bess.) Fisch. et Mey. ex Nakai.

\+ Fertile shoots with crispate pubescence. Terminal leaf lobules lanceolate or ovoid-lanceolate. Anthodia oblong .... 19.

19. Fertile shoots single, angular-sulcate. Leaf blade profile oblong-ovoid, green above, with whitish tomentum underneath. Anthodia campanulate .... 63. A. verbenacea (Kom.) Kitag.

\+ Fertile shoots 1–5, somewhat sulcate. Leaf blade profile orbicular or broad-ovoid, with whitish tomentose pubescence on both sides. Anthodia campanulate ...... 57. A. superba Pamp.

20 (11). Terminal leaf lobules of middle and upper cauline leaves relatively broad 2–10 mm) .... 21.

\+ Terminal leaf lobules of middle and upper cauline leaves narrow, 0.5–2 (3) mm broad .... 45. A. roxburgiana Wall. ex Bess.

21. Fertile shoots 80–120 cm tall, angular-sulcate, light brown. Leaf blade compactly punctate-glandular on both sides .... 8. A. argyi Levl. et Vaniot.

\+ Fertile shoots up to 17 cm tall, ribbed, dark brown to violet. Leaf blade without punctate glandules .... 22.

22. Rhizome long, nodose. Fertile shoots with cobwebby pubescence. Anthodia broad-campanulate, 3–4 mm long, single or form loose racemose inflorescence. Pancile narrow-pyramidal .... 33. A. leucophylla (Turcz. ex Bess.) Pamp.

\+ Rhizome short, not nodose. Fertile shoots pubescent with crispate hairs. Anthodia ovoid or ovoid-campanulate, 1.5–2 mm long, single or gathered into spicate inflorescence. Panicle broad-pyramidal .... 39. A. moorcroftiana Wall. ex DC.

23 (2). Plants green, glabrous or greyish green (specially leaves pubescent on both sides or only underneath with short adherent simple hairs or tomentose pubescence) .... 24.

\+ Plants greyish or brownish, compactly pubescent with long, silky hairs .... 46.

24. Lower cauline leaves long-petiolate; leaf petiole cristate, as long as or longer than leaf blade, with pinnatisected auricles at base; leaf blade with many linear or lanceolate lobules serratodentate

along margin; very similar lobules on common portion of petiole between primary lobules .... 25.

+ Lower cauline leaves short-petiolate; leaf petiole not cristate, shorter than leaf blade, with tri- or palmatisected auricles at base; leaf blade with rather few lyrate-, ternate- or palmatisected entire lobules .... 41.

25. Plants with short or long rhizome. Fertile shoots single or 3–5 together, herbaceous or somewhat lignifying at base. Inflorescence, a racemose or narrow-pyramidal, slightly leafy panicle .... 26.

+ Plants with thick, woody root. Fertile shoots many, lignifying at base. Inflorescence, a pyramidal compactly leafy panicle ...... 33.

26. Fertile shoots and leaf blade green on both sides, glabrous, rarely diffusely pubescent only at the beginning of vegetation ... . 27.

+ Fertile shoots glabrous in lower portion, pubescent in upper portion. Leaf blade compactly pubescent only on underside or on both sides .... 29.

27. Blades of lower cauline leaves oval or orbicular-oval in profile; terminal leaf lobules acutiserrate-dentate along margin; bracts longer than anthodia .... 36. A. maximovicziana Krasch. ex Poljak.

+ Blades of lower cauline leaves oblong-elliptical to broad-elliptical in profile; terminal leaf lobules sparsely dentate along margin or entire; bracts not longer than anthodia .... 28.

28. Lower cauline leaves on petioles as long as leaf blade or slightly shorter; primary leaf lobules oblong-elliptical. Anthodia on pubescent stalks .... 25. A. grubovii Filat.

+ Lower cauline leaves on petioles longer than leaf blade; primary leaf lobules lanceolate or ovoid-lanceolate. Anthodia on glabrous stalks .... 32. A. latifolia Ledeb.

29. Rhizome nodose, issuing out clusters of radical leaves apart from shoots. Fertile shoots 2–5. Inflorescence, a narrow-pyramidal panicle .... 30.

+ Rhizome stands obliquely erect, issuing out only several fertile shoots. Inflorescence, a racemose or spicate panicle .... 32.

30. Leaf blades of lower and middle cauline leaves green on top, rarely pubescent, greyish underneath due to fairly compact pubescence of long simple hairs; primary leaf lobules oblong-elliptical, stand obliquely erect .... 59. A. tanacetifolia L.

+ Leaf blades of lower and middle leaves green on both sides, glabrous or diffusely pubescent underneath; primary leaf lobules oblong, declinate almost at right angle to petiole ...... 31.

31. Plant 50–90 cm tall. Blades of lower and middle cauline leaves as long as petiole. Anthodia 2–3 mm broad. Phyllary with narrow transparent margin .... 31. A. laciniata Willd.

+ Plants up to 50 cm tall. Blades of lower and middle cauline leaves shorter than petiole. Anthodia 4–6 mm broad. Phyllary with broad dark brown margin .... 4. A. phaeolepis Krasch.

32. Blades of lower cauline leaves ovoid or broad-ovoid in profiles; primary leaf lobules elliptical or oblong-elliptical. Anthodia 3–4 mm broad, form compact spicate panicle .... 30. A. junatovii Filat.

60 + Blades of lower cauline leaves oblong in profile; primary leaf lobules ovoid-lanceolate. Anthodia 7–8 mm broad, form loose racemose panicle .... 19. A. disjuncta Krasch.

33 (25). Entire plant, specially leaves on both sides and phyllary whitish, with compact cobwebby pubescence. Anthodia ovoid. Panicle narrow-pyramidal .... 37. A. messerschmidtiana Bess.

+ Entire plant green, somewhat pubescent. Leaf blade green above, greyish underneath, fine-tomentose. Phyllary glabrous or diffusely pubescent. Anthodia hemispherical. Panicle broad-pyramidal .... 34.

34. Fertile shoots slender, rod-shaped, somewhat branched. Lower cauline leaves 2–3-pinnatisected; terminal leaf lobules closely proximated .... 21. A. freyniana (Pamp.) Krasch.

+ Fertile shoots thickish, highly branched. Lower cauline leaves bipinnatisected; terminal leaf lobules spaced or not closely proximated .... 35.

35. Terminal leaf lobules of lower and middle cauline leaves narrow-linear, lanceolate, or linear-lanceolate, elliptical, up to 1 mm long .... 36.

+ Terminal leaf lobules of lower and middle cauline leaves fili-form-linear, 4–6 mm long .... 40.

36. Blades of lower cauline leaves oblong-ovoid in profile, 3 (4)–15 cm long; terminal leaf lobules lanceolate or linear-lanceolate, acuminate. Anthodia up to 3 mm broad .... 24. A. gmelinii Web. ex Stechm.

+ Blades of lower cauline leaves ovoid or orbicular-ovoid in profile, 2–4 cm long; terminal leaf lobules narrow-linear or elliptical, subobtuse. Anthodia 6–8 mm broad .... 37.

37. Fertile shoots dirty violet, with uniform finely cobwebby pubescence. Lower cauline leaves on petioles as long as leaf blade or longer. Anthodia 4–5 mm broad .... 65. A. vestita wall. ex DC.

\+ Fertile shoots brown or dark brown, glabrous or diffusely pubescent. Lower cauline leaves on petioles shorter than leaf blade. Anthodia 5–8 mm broad .... 38.

38. Fertile shoots compactly leafy. Terminal leaf lobules linear or lanceolate. Panicle narrow-pyramidal .... 39.

\+ Fertile shoots somewhat leafy. Terminal leaf lobules elliptical. Panicle broad-pyramidal .... 3. A. aksaiensis Y.R. Ling.

39. Perennial with decumbent rhizome. Fertile shoots single or more. Lower and middle cauline leaves with diffuse pubescence of fine appressed hairs above; whitish, compactly pilose underneath .... 34. A. macrantha Ledeb.

\+ Subshrubs with stout woody root. Fertile shoots many. Lower and middle cauline leaves glabrous above, punctate-cellular; whitish, compactly pilose underneath .... 49. A. santolinifolia (Turcz. Ex Pamp.) Krasch.

40. Lower and middle cauline leaves on petioles as long as leaf blade or shorter. Anthodia 2–4 mm long, highly proximated on lateral branchlets in narrow-pyramidal panicles .... 2. A. adamsii Bess.

\+ Lower and middle cauline leaves on petioles longer than leaf blade or shorter. Anthodia (3) 4–5 mm long, spaced on lateral branchlets in broad-pyramidal panicles .... 14. A. brachyloba Franch.

41 (24). Fertile shoots single or more (3–8). Blades of lower cauline leaves oblong-ovoid in profile. Outer phyllary longer than inner .... 42.

\+ Fertile shoots many. Blades of lower cauline leaves orbicular or ovoid in profile. Outer leaflets as long as inner .... 43.

42. Fertile shoots 10–25 cm tall. Leaf blades rugose, punctate-glandular, 1.5–2 cm long. Anthodia 6–8 mm broad; phyllary linear or linear-lanceolate, with brownish scarious margins .... 67. A. viridis Willd.

\+ Fertile shoots 20–50 cm tall. Leaf blades glabrous, green, smooth, 5 cm long. Anthodia 4–7 mm broad; phyllary elliptical, with transparent scarious margin .... 47. A. rupestris L.

43. Entire plant glabrous, green. Blades of lower and middle cauline leaves 4–8 mm long. Anthodia gathered in loose

racemose inflorescences in pyramidal panicles on lateral branchlets .... 23. A. glabella Kar. et Kir.

+ Entire plant pubescent with long simple hairs, specially at commencement of vegetation. Blades of lower and middle cauline leaves 6–20 mm long. Anthodia single or 2–3 together in narrow racemose or spicate panicles on lateral branchlets ......
.... 44.

44. Rhizome aslant with many funiform roots. Fertile shoots rod-shaped, Anthodia oviod, 2.5–4 mm broad; outer phyllary broad-ovoid .... 18. A. dalai-lamae Krasch.

+ Root rachiform, thick, woody. Fertile shoots thickish, woody. Anthodia globose, 3–6 mm broad; outer phyllary oblong .... 45.

45. Fertile shoots 30–40 cm tall. Inflorescence, a narrow-pyramidal panicle. Anthodia 3–4 mm broad. Outer phyllary narrow-scarious and pilose-fimbriate along margin ....
.... 41. A. obtusiloba Ledeb.

+ Fertile shoots 5–20 (30) cm tall. Inflorescence, a racemose panicle. Anthodia 5–6 mm broad. Outer phyllary broad-scarious along margin .... 4. A. altaensis Krasch.

46 (23). Fertile shoot single, highly sulcate, 60–100 cm tall. Lower cauline leaves 6–9 cm long, 3–7 cm broad; leaf blade profile broad-oval .... 1. A. absinthium L.

+ Fertile shoots not sulcate, up to 60 (70) cm tall. Lower cauline leaves smaller .... 47.

47. Root rachiform, stout, woody. Fertile shoots many, compactly leafy .... 48.

+ Rhizome slender, decumbent. Fertile shoots single or more (3–8), somewhat leafy .... 50.

62 48. Plant up to 20 cm tall. Blades of lower cauline leaves oblong-linear in profile. Anthodia 2–3 mm broad .... 49.

+ Plant 30–60 cm tall. Blades of lower cauline leaves orbicular to ovoid in profile. Anthodia 3–6 mm broad .... 55.

49. Fertile shoots branched from base. Lateral branchlets of panicle 5–7 (9) cm long, subhorizontally declinate. Anthodia declinate, often nutant; outer phyllary ovoid ....
.... 29. A. jounghusbandii Drumm. ex Pamp.

+ Fertile shoots branched up to centre. Lateral branchlets of panicle up to 5 cm long, stand obliquely erect. Anthodia not declinate, upright, not nutant; outer phyllary linear-oblong ....
.... 15. A caespitosa Ledeb.

50. Fertile shoots highly lignifying. Blades of lower cauline leaves orbicular or reniform in profile, bipinnatisected; terminal leaf lobules linear or linear-oblong .... 51.

\+ Fertile shoots somewhat lignifying. Blades of lower cauline leaves ovoid in profile, 2-3-pinnatisected; terminal leaf lobules lanceolate or linear-lanceolate .... 56.

51. All leaves (except bracts) petiolate, petiole without lobules at base; terminal leaf lobules 2 pairs, subobtuse at tip. Phyllary white-tomentose .... 52.

\+ Lower cauline leaves petiolate, rest sessile, or all leaves sessile, with lobules at base; terminal leaf lobules 2–3 (5) pairs, acuminate at tip. Phyllary greyish-pilose .... 53.

52. Fertile shoots light green. Leaf blade of lower and upper cauline leaves tripinnatisected. Terminal leaf lobules lanceolate-linear. Outer phyllary greyish-cobwebby .... 43. A. persica Boiss.

\+ Fertile shoots brownish grey. Leaf blades of lower and upper cauline leaves bipinnatisected. Terminal leaf lobules linear-oblong, subobtuse. Outer phyllary oblong, white-tomentose .... 48. A. rutifolia Steph. ex Spreng.

53. Fertile shoots thick, rough, many-branched, highly lignifying, light brown. Terminal leaf lobules closely proximated, 0.5–1.5 mm long. Panicle loose, broad-pyramidal .... 69. A. xerophytica Krasch.

\+ Fertile shoots slender, lignifying in lower part, branched up to centre or only in upper portion. Terminal leaf lobules uniformly spaced, 2–5 mm long. Inflorescence, a racemose or narrow-pyramidal panicle .... 54.

54. All leaves sessile, 0.5–1.2 cm long. Anthodia 5–6 mm broad, form contracted, loose racemose panicle. Inner phyllary elliptical, with broad dark brown margin .... 9. A. argyrophylla Ledeb.

\+ Lower cauline leaves petiolate, 1–2 cm long, rest sessile. Anthodia 3–4 mm broad, form compact, contracted, narrow-pyramidal panicle. Inner phyllary oblong-elliptical, with narrow transparent margin .... 22. A. frigida Willd.

55. Lower cauline leaves on petioles considerably longer than leaf blade, rest sessile, whitish on both sides, compactly pubescent with simple hairs. Anthodia globose, stalked .... 54. A. stracheyi Hook. f. et Thoms. ex Clarke.

+ Lower cauline leaves on petioles as long as leaf blade or shorter, rest-petiolate, with compact brownish tomentum on both sides. Anthodia broad-ovoid, sessile ....
.... 16. A. campbellii Hook. f. et Thoms. ex Clarke.

56 (50). Blades of lower cauline leaves ovoid in profile. Anthodia broad-ovoid, up to 2 mm broad, form pyramidal panicle ....
.... 12. A. austriaca Jacq.

+ Blades of lower cauline leaves broad-ovoid or reniform in profile. Anthodia hemispherical, 4–7 mm broad, form racemose panicle .... 57.

57. Fertile shoots 1–3. Terminal leaf lobules sessile, linear or linear-lanceolate, 10–20 mm long. Anthodia 4–6 mm broad; outer phyllary oblong-ovoid, acuminate ....
.... 51 A. sericea Web. ex Stechm.

+ Fertile shoots quite many (15–20). Terminal leaf lobules petiolate, lanceolate, 8–10 mm long. Anthodia 5–7 mm broad; outer phyllary oblong, obtuse ....
.... 10. A. aschurbajewii Winkl.

58 (1). At the commencement of vegetation, plants rather sparsely pubescent with simple, adherent hairs, later glabrous, green .... 59.

+ Plants greyish or whitish up to end of vegetation due to compact tomentose pubescence .... 64.

59. Terminal leaf lobules narrow-linear or filiform .... 60.

+ Terminal leaf lobules broader .... 61.

60. Blades of lower cauline leaves 2–5 cm long, bipinnatisected; terminal leaf lobules filiform. Anthodia aggregated into compact, contracted, spicate, and racemose inflorescences on lateral branchlets and form narrow, compressed panicle ....
.... 42. A. palustris L.

+ Blades of lower cauline leaves 6–12 cm long, bipinnatisected; terminal leaf lobules narrow-linear. Anthodia in compact globose glomerules on lateral branchlets and form broad-pyramidal panicle .... 11. A. aurata Kom.

61. Fertile shoots green or brownish green. Inflorescence, a broad diffuse panicle. Anthodia declinate, nutant .... 7. A. annua L.

+ Fertile shoots dark brown or brownish violet. Inflorescence, a narrow, compact panicle. Anthodia upright or subdeclinate, not nutant .... 62.

62. Annual-biennial plant 120–200 cm tall. Lower and middle cauline leaves short-petiolate. Phyllary green, lustrous ....
.... 62. A. tournefortiana Reichb.

+ Annual plant 10–45 cm tall. All leaves sessile. Phyllary blackish, dark brown, or brownish along margin .... 63.

63. Fertile shoots branched from base. Panicle narrow-pyramidal, with short lateral branchlets. Phyllary dark brown or blackish along margin .... 26. A. hedenii Ostenf.

+ Fertile shoots branched from centre. Panicle broad-pyramidal, with long lateral branchlets. Phyllary brownish along margin, nearly transparent .... 17. A. chamomilla Winkl.

64. Fertile shoots erect, branched from base. Blades of lower cauline leaves oblong in profile. Anthodia ovoid, up to 2 mm long, phyllary fimbriate along margin ....
.... 13. A. blepharolepis Bge.

+ Fertile shoots branched from centre. Blades of lower cauline leaves broad-ovoid to broad-deltoid in profile. Anthodia 2–10 mm long; phyllary without cilia along margin .... 65.

65. Blades of lower cauline leaves ovoid or oblong-ovoid in profile. Anthodia 5–10 mm broad, long-stalked, gathered in racemose inflorescence on lateral branchlets and form loose broad-pyramidal panicle .... 35. A. macrocephala Jacq. ex Bess.

+ Blades of lower cauline leaves broad-ovoid or broad-deltoid in profile. Anthodia 2–5 (6) mm broad, on very short stalks or sessile, gathered in groups of 2–3 or singly on lateral branchlets to form broad panicle .... 66.

66. Fertile shoots stout, single or 2–3 together, highly ribbed. Blades of lower and middle cauline leaves broad-deltoid in profile; terminal leaf lobules linear or linear-oblong, obtuse at tip. Anthodia 4–5 mm broad .... 67.

+ Fertile shoots slender, many, not ribbed. Blades of lower and middle cauline leaves broad-ovoid in profile; terminal leaf lobules filiform, acuminate at tip. Anthodia 2–4 mm broad ....
.... 68.

67. Lower cauline leaves on petioles as long as leaf blade or longer, 2–3-pinnatisected. Outer phyllary linear-oblong ....
.... 52 A. sieversiana Willd.

+ Lower cauline leaves on petioles shorter than leaf blade, 1–2-pinnatisected. Outer phyllary elliptical or ovoid ....
.... 56. A. succulentoides Ling et Y.R. Ling.

68. Lower cauline leaves 1.5–5 cm long. Terminal leaf lobules filiform, 10–15 mm long, geniculately curved, proximated. Anthodia 2–4 (6) mm broad ....
.... 5. A. anethifolia Web. ex Stechm.

+ Blades of lower cauline leaves up to 1.5 cm long. Terminal leaf lobules filiform, up to 4 mm long, erect, not curved, spaced. Anthodia 2–2.5 mm broad .... 6. A. anethoides Mattf.

1. A absinthium L. Sp. pl. (1753) 843; Krasch. in Kryl. Fl. Zap. Sib. 11 (1949) 2817; Poljak. in Fl. SSSR, 26 (1961) 515; Fl. Sin 76, 2 (1991) 12; Fl. Kirgiz.11 (1965) 172; Fl. Kazakhst. 9 (1966) 102; Opred. rast. Sr. Azii [Key to Plants of Mid. Asia] 10 (1993) 557.

Described from Europe. Type in London (Linn.).

Weed on roadsides and in residences, gardens, kitchen gardens, plantations, pastures.

IIA. Junggar: *Cis-Alt.* (15 km north-west of Shara-Sume, in plantations, Aug. 7, 1959 – Yun et Yuan'), *Tien Shan* (Boro-Khoro mountain range, 30 km from Chingi-khodzy settlement on road to Urumchi from Kul'dzha, scrubs, Aug. 19. 1957 – Yun., Li et Yuan').

General distribution: Aralo-Casp., Fore Balkh., Jung.-Tarb., North and Cent. Tien Shan, West and East. Sib., Far East, China (North), Himalayas, North America.

65

2. A. adamsii Bess. in Nouv. Mem. Soc. natur. Moscou, 3 (1834) 262; Kitag. Lin. Fl. Mansh. (1939) 422; Grub. Konsp. fl MNR [Conspectus of Flora of Mongolian People's Republic] (1955) 262; Poljak. in Fl. SSSR, 26 (1961) 463; Leonova in Grub. Opred. rast. Mong. [Key to Plants of Mongolia] (1982) 251; F1. Intramong. 6 (1982) 143; F1. Sin. 76, 2 (1991) 69; Fl. Intramong. ed. 2, 4 (1993) 619; Gub. Konsp. fl. Vneshn. Mong. [Conspectus of Flora of Outer Mongolia] (1996) 98. — *A. licentii* Pamp. in Nouv. Giorn. Bot. Ital. n.s., 34 (1927) 676. — Ic.: Fl. SSSR, Plate 24, fig. 1; Grub. Opred. rast. Mong. [Key to Plants of Mongolia] (1982), Plate 82, fig. 604; Fl. Intramong. (1982), tab. 51, fig 7, 8; Fl. Intramong. (1993), tab. 243, fig. 7, 8.

Described from East. Sib. Type in Kiev (KW).

On steppe, rocky and clayey slopes and trails of mountains and conical hillocks, solonchak and solonetzic meadows, banks of rivers and lakes, around springs, ravines, on roadsides and in residences as weed.

IA. Mongolia: *Mong. Alt.* (nor. slopes of Batyr-Khairkhan mountain range, Tasbun-obo river valley, Nov. 2, 1899 – Lad.; slopes of Botkhon gorge, Sept. 1, 1930 – Bar.; Khan-Taishiri mountain range, Sept. 3, 1948 – Grub.; Taishir-ula mountain range, 7 km south of Altay town, July 2, 1973 – Gub.), *Cent. Khalkha* (Ikhe-Tukhum-nor lake, Aug. 8; 60–80 km south-west of Urga, Aug. 8; Dzhargalant river basin, arid bed, Aug. 26–1925, Krasch. et Zam.; 100 km south-east of Ulan-Bator town, hummocky region, Sept. 6, 1950-Ivan.; Sumbere state farm, 35 km north-west of Choiren, Aug. 23, 1961 – Dalgayam), *East. Mong.* (vicinity of Dariganga, Aug. 2, 1927 – Zam.; vicinity of Dashibalbar somon, Aug. 24, 1949 – Yun.; 2 km west of Enger-Shanda, June 6, 1958 – Dalgayam; "Manchuria" – Fl. Intramong. (1993) l.c.), *Depr. Lakes* (south. bank of Khara-nur lake, 20 km west of Santa-Margats somon, Aug. 18, 1944 – Yun.), *Val. Lakes* (Ongiin-gol river, among fine rubble, July 27, 1893 – Klem.; valley of Tatsiin-gol river, July 24, 1924 – Gorbunova; 40 km south of Arbaikhere town and 35 km north of Tugrek settlement, Aug. 26, 1983 – Gub.), *East. Gobi* (40 km north of Sain-Shanda, July 17, 1940 – Yun.).

72

General distribution: East. Sib. (Sayans), Far East, North Mong. (Hent., Hang., Mong.-Daur.).

3. **A. aksaiensis** Y.R. Ling in Bull. Bot. Research, 5, 2 (1985) 3; Fl. Sin. 76, 2 (1991) 43; Fl. Intramong. ed. 2, 4 (1993) 610; Gub. Konsp. fl. Vneshn. Mong. [Conspectus of Flora of Outer Mongolia] (1996) 96. — Ic.: Fl. Intramong. (1993), tab. 241, fig. 11, 12.

Described from Mongolia (Khesi). Type in Beijing (PE). Map 6.

On rubbly and rocky slopes, among rocks.

IA. Mongolia: *Khobd., Cent. Khalkha* (Gub. l.c.), *Gobi Alt.* (Gurban-Saikhan-ula, rocky slopes, Aug. 27, 1970—Santar), *East. Gobi* (200 km south of Sain-Shanda town, Barun-Nariin-ula mountains, rubbly slopes, 1100 m, July 8; same site, Khutang-ula mountains, among rocks, July 10—1982, Gub.), *Alash. Gobi* (Alashan mountain range, Yamata river gorge, July 13, 1908—Czet.), *Ordos* (valley of Ulan-Marin river, Aug. 22, 1884—Pot.), *Khesi* ("vicinity of Akai settlement"–typus!).

General distribution: Nor. Mong. (Heng.).

4. **A. altaensis** Krasch. in Kryl. Fl. Zap. Sib. 11 (1949) 2792; Grub. Konsp. fl. MNR [Conspectus of Flora of Mongolian People's Republic] (1955) 262; Poljak. in Fl. SSSR, 26 (1961) 514; Leonova in Grub, Opred. rast. Mong. [Key to Plants of Mongolia] (1982) 250; Gub. Konsp. fl. Vneshn. Mong. [Conspectus of Flora of Outer Mongolia] (1996) 96.— *A. obtusiloba* Ledeb. var. *fruticosa* Ledeb. Fl. Alt. 4 (1833) 69. — Ic: Ic. pl. Fl. ross. 5 (1833), tab. 466.

Described from Siberia (Altay). Type in St.-Petersburg (LE).

On rubbly and rocky desert-steppe slopes of mountains, on rocks.

IA. Mongolia: *Khobd.* (Khoiligin-daba, alpine desert steppe, Aug. 3, 1930-Bar.; Ulyungin-daba, upper portion of nor. slope to pass, Aug. 3; 4 km westward of Sagil somon, Ulan-daba pass, mountain steppe, July 29-1945, Yun.; conical hillock 16 km south of Tsagan-nur somon, sheep's fescue steppe, Aug. 24, 1977—Karam. et Sanczir), *Mong. Alt.* (Shar'ya-Mally river valley, steppe slopes, July 27, 1905—Sap.; upper course of Unchin-gol river, alpine desert steppe, Sept. 12, 1930—Bar.; Tolbo-Kungei mountain range, arid steppe, Aug. 5, 1945—Yun.; nor. bank of Dayan-nur lake, coastal strip, Aug. 8, 1972—Metel'tseva).

General distribution: West. Sib. (Altay), Nor. Mong. (Hang.).

5. **A. anethifolia** Web. ex Stechm. Artemis (1775) 29; Kitag. Lin. Fl. Mansh. (1939) 422; Grub. Konsp. fl. MNR [Conspectus of Flora of Mongolian People's Republic] (1955) 262; Poljak. in Fl SSSR, 26 (1961) 522; Leonova in Grub. Opred. rast. Mong. [Key to Plants of Mongolia] (1982) 262; Fl. Intramong. 6 (1982) 153; Pl. vasc. Helansch. (1986) 356; Fl. Sin. 76, 2 (1991) 32; Fl. Intramong. ed. 2, 4 (1993) 601; Gub. Konsp. fl. Vneshn. Mong. [Conspectus of Flora of Outer Mongolia] (1996) 96. — Ic.: Grub. Opred. rast. Mong. [Key to Plants of Mongolia] Plate 131, fig. 60; Fl. Intramong. (1982) tab. 54, fig. 8, 9; Fl. Intramong. (1993) tab. 236, fig. 7, 8.

Described from East. Siberia. Type in St.-Petersburg (LE).

In desert zone, steppified deserts on solonchaks, chee grass thickets, on banks of rivers, lakes, and irrigation canals.

IA. Mongolia: *Cent. Khalkha* (Tukhum-nur lake, on arid solon-chak, Aug. 3, 1893 — Klem.; midcourse of Uber-Dzhergalante river, near Dol' che-Gegen monastery, solonchak, Aug. 25, 1925 — Krasch. et Zam.; vicinity of Ikhe-Tukhum-nur lake, Tashiga mountain, June 1926 — Polynov et Lebedev; floodplain of Kerulen river, 40 km nor. — east to eastern part of Bayan-Berkhen-obo mountain, solonchak, Aug. 8, 1970 — Mirkin et al), *East. Mong.* (southward of Dalai-nur lake, 1899 — Pal.; Khalkhin-gol, 30 km from Bayan-Burdu, alkali grass meadow, Aug. 18; Shavorte-nur lake, 50 km east of Erentsab, solonchak meadow, Aug. 19; same site, Shavorte-obo area, in ravine, Aug. 20; Dashi-Balbar, Khargaitu area, bent grass meadow, Aug. 22–1949. Yun.; 45 km north-west of Khailar town, Sept. 12, 1958 — Lavr.; Ongon-Elisu sand, southward of Khongor settlement, July 31, 1981 — Gub.), *Depr. Lakes* (Khara-Us-nur lake, Aug. 17, 1879 — Pot.; Shargain-gol river, between Shargain-Tsagan-nur lake and Dzak-Obo, solonchak, Sept. 8, 1930 — Pob.; Kobdo river valley, on Kobdo-Ulangom road, Aug. 23, 1944 — Yun.), *Val. Lakes* (Tatsiin-gol river, July 17, 1893 — Klem.), *Gobi Alt.* (Dzolin-ula, 40 km eastward of Sevrei settlement, Aug. 26, 1982 — Gub.), *East. Gobi* (Sain-usu sand, Aug. 20, 1926 — Lis.; Alkha-Khoshuin-Gobi, on terrace, Aug. 14, 1927 — Zam.; Sain-Tsagan, 37 km southward of Mandal-Gobi, near saline lake, Aug. 12, 1950 — Lavr.; hummocky area south — south-east of Khan-Bogdo somon, 10 km north-west of Bosgony-khuduk well, in gorge, July 20; 70 km south — south-west of Khatan-Bulak somon, desert plain, July 25; 75 km north-west of Sain-Shanda on road to Khara-airik, solonchak around Ulan-nur lake, Aug. 31–1974, Rachk. et Volk.; Khan-Bogdo-ula, 20 km south-east of Khan-Bogdo settlement, Sept. 2, 1982 — Gub.), *West. Gobi* (Atas-Bogdo-ula, Aug. 2, 1978 — Gub.).

IB. Kashgar: *East.* (near Shuguz well, floor of arid bed, Aug. 11, 1895 — Rob.).

IIA. Junggar: *Jung. Gobi* (105 km south — south-east of Bulgan somon, saxaul thicket, Aug. 11, 1977 — Rachk. et Volk.).

General distribution: West. Sib. (Altay), East. Sib. (Sayans), Far East, North Mong. (Hang., Mong.-Daur.), China (North.-common).

6. A. anethoides Mattf. in Feddes Repert. 32 (1926) 249; Kitag. Lin. Fl. Mansh (1939) 423; Fl. Intramong. 6 (1982) 153; Pl. vasc. Helansch. (1986) 357; Fl. Sin. 76, 2 (1991) 33; Fl. Intramong. ed. 2, 4 (1993) 601; Gub. Konsp. fl. Vneshn. Mong. [Conspectus of Flora of Outer Mongolia] (1996) 96. — Ic.: Fl. Intramong. (1982) tab. 54, fig. 6, 7; Fl. Intramong. (1993) tab. 236, fig. 9, 10.

Described from East. China. Type in Berlin (B).

On saline clayey soil and saline sand in desert zone, in river valleys, ravines, on banks of lakes.

IA. Mongolia: *East. Mong.* (Kuku-Khoto, plain north-west of Guikhua-in', July 31; same site, northward of Mekou town, Aug. 3, 1884; same site, Udzhyum river valley, Aug. 16–1886, Pot.; Dabasutai-nor lake, on bank, 1899 — Pal.; Khotun Ikhe-Dulan, plain, Aug. 19, 1928 — Lebedev; foothills of Ikhe-Dulan mountains, saline soil, Aug. 8; Ulan-Khairkhan rise, Aug. 27–1928, Shastin; vicinity of Baga-ude collective, Aug. 21; Dariganga, vicinity of Argulsul mountains, on takyr (clay-surfaced desert), Aug. 31; same site, on road to Kholu-Marte well toward Koshun-Shent well, solonchak, Sept. 7–1931, Pob.; Urgul, 160 km from Baishintu on road to Sain-Shandu, Aug. 26, 1950 — Ivan.; Dariganga, vicinity of dried up lake, Aug. 25, 1927 — Zam.; 100 km south-east of

Choibalsan, solonchak lowland, Aug. 27, 1954 – Dashnyam; Shilin-Khoto town, 1959 –
Ivan.), *Val. Lakes* (Orok-nur lake, saline sand, Aug. 6, 1926 – Tug.), *East. Gobi* (Baga-
Ude, ravine in Khara-Ula mountain, Aug. 16, 1926 – Lis.), *Gobi Alt.* (Bayan-Tsagan
mountain range, Khabtsyn area, on slope, Aug. 28, 1886 – Pot.; foothills of Ikhe-Bogdo
mountain range, Aug. 24, 1926 – Tug.; Bayan-Tsagan mountain range, rocky slopes of
mounds, Sept. 3, 1927 – Simukova; Bayan-Tukhum area, sand along gorges, Sept. 1;
Tszolin mountains, among boulders, Sept. 13-1931, Ik.-Gal.; Ikhe-Bogdo mountain
range, desert steppe, Sept. 13, 1943 – Yun.), *East. Gobi* (Sain-Usu, sand, Aug. 20, 1926-
Lis.; Dalan-Dzhirgalan somon, Delgeriin-Deris area, 40 km south-east of somon, saline
derris thicket (tropical woody climber), Aug. 24; Khara-Airik mountain range, 10 km
north-east of somon on road to Choiren, solonchak barren land, Aug. 26; west. extremity
of Sain-Usu basin, Aug. 28; 16 km north-east of Sain-Shanda on road to Baishintu, derris
thickets, Sept. 1; Abdarantin-tsob, 2 km south of Sain-Shanda, floor of gorge, Sept. 4;
Barun-Tugrig area, 30 km west of Sain-Shanda, solonchak, Sept. 7; Delgirkh somon,
Khotungei area, floor of lowland, Sept. 11-1940, Yun.; central portion of Khurkhool
mountains, in gorge, July 15; Galbyn-Gobi desert, 45 km south-east of Khan-Bogdo, in
saxaul thicket, July 20; 150 km south – south-east of Sain-Shanda, Aug. 4; 35 km east of
Under-Shili, solonchak, Aug. 10-1974, Rachk. et Volk.), *West. Gobi* (10–15 km westward
of Tsagan-Bulak somon, near southern foothills of Tsagan-Bogdo mountain range, along
floor of gorge, Aug. 5, 1943 – Yun.; same site, in ravine, Aug. 27, 1979 – Grub., Muld.,
Dar.), *Ordos* (Ordos, valley of Huang He river, Aug. 23; same site, Aug. 24 – 1871,
Przew.; Ordos, 1884 – Bretshneider; Chagon-nor lake, Aug. 24, 1884 – Pot.), *Khesi* (40 km
east of Chzhan'e town, on arid bed, Aug. 10, 1958 – Petr.).

IIIB. Tibet: *Weitzan* (Burkhan-Budda area, April 17-29, 1884 – Przew.).

General distribution: China (Nor., East.–common).

7. **A. annua** L. Sp. pl. (1753) 847; Franch. Pl. David. (1884) 169; Forbes
et Hemsl. Index Fl. Sin. 1 (1888) 441; Kitag. Lin. Fl. Mansh. (1939) 423;
Krasch. in Kryl. Fl. Zap. Sib. 11 (1949) 2816; Grub. Konsp. fl. MNR
[Conspectus of Flora of Mongolian People's Republic] (1955) 263; Poljak.
in Fl. SSSR, 26 (1961) 489; Fl. Kirgiz. 11 (1965) 164; Fl. Kazakhst. 9 (1966)
95; Enum. vasc. pl. Xizang. (1980) 361; Leonova in Grub. Opred. rast.
Mong. [Key to Plants of Mongolia] (1982) 247; Fl. Intramong. 6 (1982)
146; Fl. Xizang. 4 (1985) 754; Pl. vasc. Helansch. (1986) 257; Fl. Sin. 76, 2
(1991) 62; Fl. Intramong. ed. 2, 4 (1993) 618; Opred. rast. Sr. Azii [Key to
Plants of Mid. Asia] 10 (1993) 553; Gub. Konsp. fl. Vneshn. Mong.
[Conspectus of Flora of Outer Mongolia] (1996) 96. — Ic.: Amman, Stirp.
rar. tab. 193, fig. 23; Gmel. Fl. Sib. 2 (1845) tab. 125; Fl. Intramong. (1982)
tab. 52, fig. 9, 10; Fl. Xizang. (1985) tab. 324, fig. 1–5; Fl. Intramong. (1993)
tab. 242, fig. 9, 10.

Described from Europe. The sketch of Amman and Gmelin serves as
type.

Weed around residences, roadsides, in farms, gardens, kitchen
gardens, near wells, on pasture corrals and in wintering sites, on banks
of rivers and lakes, floors of gorges.

IA. Mongolia: *Mong. Alt., Cent. Khalkha, East. Mong., Depr. Lakes, East. Gobi, Gobi
Alt., West. Gobi, Alasch. Gobi, Ordos.*

IB. Kashgar: *Nor.* (between Aksu and Kucha, Aug. 15, 1929 — Pop.), *East.* (Lyukchun depression, Sept. 27, 1895 — Rob.).

IIA. Junggar: *Jung. Gobi* (Bulun-Tokhoi, Aug. 5–16, 1876 — Pot.).

IIIA. Qinghai: *Nanshan* (Tangug, Aug. 1872; South Tetungsk mountain range, July 28–1880, Przew.; Van-tszy-tszin', on bank of irrigation canal, July 16, 1908 — Czet.).

IIIB. Tibet: *South.* ("Lhasa" — Fl. Xizang. l.c.).

General distribution: Aralo-Casp., Fore Balkh., Jung.-Tarb., Nor. and Cent. Tien Shan, East. Pam., Europe, Mediterr., Balk.-Asia Minor, Caucasus, Mid. Asia, West. and East. Sib., Nor. Mong., China, Himalayas, Japan, North America.

8. A. argyi Levl. et Vaniot in Feddes Repert. 8 (1910) 138; Kitag. Lin. Fl. Mansh. (1939) 423; Grub. Konsp. fl. MNR [Conspectus of Flora of Mongolian People's Republic] (1955) 262; Poljak. in Fl. SSSR, 26 (1961) 451; Leonova in Grub. Opred. rast. Mong. [Key to Plants of Mongolia] (1982) 242; Fl. Intramong. 6 (1982) 140; Fl. Sin. 76, 2 (1991) 87; Fl. Intramong. ed. 2, 4 (1993) 623; Gub. Konsp. fl. Vneshn. Mong. [Conspectus of Flora of Outer Mongolia] (1996) 96. — Ic.: Fl. Intramong. (1982) tab. 46, fig. 1–8; Fl. Intramong. (1993) tab. 244, fig. 1–8.

Described from China, Type in Paris (P).

Borders of larch forests, forest glades and meadows, among shrubs, rarely in fallow lands.

IA. Mongolia: *Cent. Khalkha* (Munon-Ula mountains, rocky slopes, Oct. 3, 1985 — Kam. et Dar.), *East. Mong.* (East. Mong., 1841–Kirilov; between Ulan-Khod and Tsindai, Aug. 10, 1898 — Zab.; Dariganga, Ongon-Elisu sand, near Boro-Bulak spring, Oct. 13, 1931 — Pob.; Shilin-Khoto, feather grass-wild rye steppe, 1959 — Ivan.; vicinity of Shimin-Bogdo-ula mountains, Aug. 7, 1964 — Dashnyam, "Shilin-Khoto, Khukh-Khoto [Datsinshan']" — Fl. Intramong. (1993) l.c.), *Ordos* (east. — Fl. Intramong. (1993) l.c.).

General distribution: Far East, Nor. Mong. (Hent., Mong.-Daur., Fore Hing.), China (Nor., East.).

Note. Closely related to *A. leucophylla* (Turcz. ex Bess.) Pamp. differing from it in much higher fertile shoots (up to 120 cm), and lobules of lower and middle cauline leaves with entire or short-toothed tip, and narrower, almost oblong anthodia.

9. A. argyrophylla Ledeb. Fl. Alt. 4 (1833) 66; Krasch. in Kryl. Fl. Zap. Sib. 11 (1949) 2799; Grub. Konsp. fl. MNR [Conspectus of Flora of Mongolian People's Republic] (1955) 263; Leonova in Grub. Opred. rast. Mong. [Key to Plants of Mongolia] (1982) 250; Fl. Sin. 76, 2 (1991) 16; Fl. Intramong. ed. 2, 4 (1993) 604; Gub. Konsp. fl. Vneshn. Mong. [Conspectus of Flora of Outer Mongolia] (1993) 96. — *A. frigida* auct. non Willd.; Poljak. in Flore SSSR, 26 (1961) 496.

Described from Siberia (Altay). Type in St.–Petersburg (LE). Map 6.

On rocky placers, old moraines in alpine mountain belt, mountain steppes and moss-lichen tundra, along banks of mountain rivers.

IA. Mongolia: *Khobd.* (Kharkhira summit, July 22, 1879 — Pot.; in pass from Uregnur basin into Achit-nur, steppe, July 30, 1945 — Yun.), *Mong. Alt.* (Taishir-ula mountains, July 17; Tsastu-Bogdo, upper course of Uzun-Dzhor river, talus, July 31; upper course of

Uinchi river, on slopes, Sept. 13–1894, Klem.; Adzhi-Bogdo mountain range, upper Indertiin-gol, mountain steppe, Aug. 6, 1947 – Yun.; Tamchi-daba pass, wormwood-sheep's fescue steppe, 2700 m, Sept. 7; 7 km south of Tonkhil'-nur lake on road to Tamchi-daba pass, mountain steppe, Sept. 7–1948, Grub.; Adzhi-Bogdo mountain range, plateau-like summit, sheep's fescue-wormwood steppe, Aug. 22, 1973 – Isach. et Rachk.; same site, 2800 m, mountain steppe, July 27; Ulan-daba pass, steppe, 2945 m, July 27, 1977 – Volk.; Ulan-daba pass, on road to Uinchi somon, steppe, 2845 m, Aug. 13, 1979 – Grub., Muld. et Dar.; 50 km south of Munkh-Khairkhan settlement, July 24; upper Bodonchiin-gol, on slopes, 2700 m, July 25–1979, Gub.; 22 km south-east of Altyn-Tsugu somon, Tokhto-ula, July 19; 13 km south-east of Bayan-Enger, Mne-daba pass, wormwood-cereal grass steppe, 2900 m, July 28–1979, Karam.; Urten-Khuren-ula town, 20 km north-east of Bulgan somon, July 4; Bulgan-gol basin, valley of Ulyasutai-gol river, midcourse, July 8, 1984 – Kam. et Dar.; 42 km south of Ulangom on road to Kobdo, wormwood-feather grass steppe, Aug. 22, 1944; 8–10 km south-east of Airag-nur lake, wormwood-crested hair grass mountain steppe, Aug. 20, 1945 – Yun.), *East. Mong.* ("Shilingol, arid steppe" – Fl. Intramong. (1993) l.c.), *Gobi Alt.* (Ikhe-Bogdo mountain range, Burgas upper creek valley, sedge-*Cobresia* meadow, Sept. 6; same site, mountain steppe, Sept. 11–1943, Yun.; same site, wormwood-cereal grass steppe, 2900 m, Aug. 4, 1973 – Isach.), *West. Gobi* (Atas-Bogdo-Ula, upper mountain belt, Aug. 12, 1943 – Yun.; same site, wormwood-wheat grass steppe, 2900 m, Aug. 22, 1976 – Rachk.).

General distribution: West. Sib. (Altay), Nor. Mong. (Fore Hubs., Hang.).

10. **A. aschurbajewii** Winkl. in Acta Horti Petrop. 11, 2 (1892) 332; Poljak. in Fl. SSSR, 26 (1961) 503; Fl. Sin. 76, 2 (1991) 12; Fl. Kirgiz. 11 (1965) 168; Fl. Kazakhst. 9 (1966) 101; Opred, rast. Sr. Azii [Key to Plants of Mid. Asia] 10 (1993) 556.

Described from Pamir. Type in St.–Petersburg (LE). Map 6.

In subalpine and alpine mountain belts, on meadow and meadow-steppe slopes and along river valleys.

IIA. Junggar: *Tien Shan* (Sairam-nur lake, July 23, 1878 – Fet.; Bogdo, 2400–2700 m, Aug. 24, 1878; Aryslan river, July 7, Aug. 10, Sept. 8–1879, A. Reg.; valley of Urtak-Sara river on road to Sairam-nuru from Borotala, steppe along valley floor, Aug. 8, 1957 – Yun., Li et Yuan'; 10 km south of Sairam-nur lake, wormwood-cereal grass steppe, Aug. 31, 1959 – Petr.).

IB. Kashgar: *south* (nor. slope of Russky mountain range, Karasai village, June 3, 1890–Rob.).

General distribution: Jung.-Tarb., Nor. and Cent. Tien Shan, East. Pamir.

11. **A. aurata** Kom. in Acta Horti Petrop. 8 (1901) 422; Poljak. in Fl. SSSR, 26 (1961) 492; Fl. Sin. 76, 2 (1991) 71; Fl. Intramong. ed. 2, 4 (1993) 622; Gub. Konsp. fl. Vneshn. Mong. [Conspectus of Flora of Outer Mongolia] (1996) 96.

Described from China (Dunbei). Type in St.–Petersburg (LE).

On steppe rocky slopes, rock crevices, rock placers.

IA. Mongolia: *East. Mong.* (Kulun-Bairnor plain, Dzyurgen-karaul town on Syel'-chzha river, in floodplain, July 27, 1899 – Pot.; vicinity of Baishintin-Sume, Ongon-Elis sand, Aug. 20, 1927 – Zam.; 150 km below Tsetsen-Khan, Aug. 12; Buir-nur lake, Aug. 27–1928, Tug.; Dariganga, 40 km north-east of Argal-Ul, Sept. 5, 1931 – Pob.; Khalkha-gol somon, valley of Khalkhin-gol river, 3 km east of Khamar-Daba, steppe, Aug. 11; vicinity of Gurban-Zagal somon, Sumen-Khevchu area, gully slope, Aug. 22–1949, Yun.;

vicinity of Khailar town, sand mounds, July 29, 1954 – Wang; same site, Sept. 13, 1958 – Lavr., same site, meadow steppe, 1959 – Ivan.; Khailar town-Fl. Intramong. (1993) l.c.), *Gobi Alt.* (Barun-Saikhan mountains, on slope among rock placers, Sept. 20, 1931; Dzun-Saikhan mountains, in depressions, Aug. 24, 1931 – Ik.-Gal.).

General distribution: Far East, Nor. Mong. (Hent., Hang., Mong.-Daur.), China (Dunbei), Korean peninsula, Japan.

Note. Closely related to *A. palustris* L., differing from it in more intricately laciniated leaf blade and broad-diffuse panicle.

12. A. austriaca Jacq. in Murr. Syst. (1784) 744; Krasch. in Kryl. Fl. Zap. Sib. 11 (1949) 2800; Poljak. in Fl. SSSR, 26 (1961) 498; Fl. Kirgiz. 11 (1965) 167; Fl. Kazakhst. 9 (1966) 100; Fl. Sin. 76, 2 (1991) 73; Fl. Intramong. ed. 2, 4 (1993) 621; Opred. rast. Sr. Azii [Key to Plants of Mid. Asia] 10 (1993) 555.

Described from Europe (Austria). Type in Wien (Vienna) (W).

On solonetzic meadows, rubbly-rocky slopes, sand, as weed around residences, roadsides, farms, fallow lands.

70   IIA. Junggar: *Balkh.-Alak.* (Sary-Khulsyn settlement, south. Durbul'dzhin, short-grass meadow on bank of brook, July 22, 1947 – Shumakov; Chuguchagsk basin 2 km westward of Durbul'dzhin on road to Chuguchak, sophora-wormwood fallow land on gently inclined trails of Tarbagatai, Aug. 7, 1957 – Yun., Li et Yuan').

General distribution: Aralo-Casp., Fore Balkh., Jung.-Tarb., Nor. Tien Shan, Europe, Mediterr., Balk.-Asia Minor, Fore Asia, Caucasus, Mid. Asia, West. Sib.

13. A. blepharolepis Bge. in Mem. Ac. Sci. Petersb. 7 (1852) 164; Grub. in Novit. syst. pl. vasc. 9 (1972) 282; Leonova in Grub. Opred. rast. Mong. [Key to Plants of Mongolia] (1982) 246; Fl. Intramong. 6 (1982) 114; Fl. Sin. 76, 2 (1991) 225; Fl. Intramong. ed. 2, 4 (1993) 654; Gub. Konsp. Fl. Vneshn. Mong. [Conspectus of Flora of Outer Mongolia] (1996) 96. – Ic.: Fl. Intramong. (1982) tab. 40, fig. 8, 9.

Described from Inner Mongolia (East. Gobi). Type in St.-Petersburg (LE).

On rather thin sand, slopes of low mountains, sandy steppes.

IA. Mongolia: *East. Mong.* ("Shilingol' ajmaq – administrative territorial unit in Mongolia – Ulantsab" – Fl. Intramong. (1993) l.c.), *Gobi Alt.* (Nemegetu-nuru mountain range, on slopes, Sept. 5, 1979 – Grub.; 45 km north-east of Bain-Tsagan, on sand among almond bushes, Aug. 4, 1983 – Gub.), *East. Gobi* (30 km north-east of Khan-Bogdo, Sept. 23, 1940-Yun.; Ulan-Tologoi, in sand desert, Aug. 1831-Bunge, typus!), *Alash. Gobi* (Alashan', near Tsagan-nur lake, Sept. 10, 1871 – Przew.; 90 km south-east of Chzhunvei town, sand-covered arid bed, June 30, 1957; Uvei, Maagantsz'sh' settlement, south. fringe of Tengeri sand, in depression between barkhan sand mounds, June 23, 1958, Zhayan-Khoto, Tengeri sand, Guntu-nor, Aug. 14, 1958; Gengeot sand, Dzhalamu, Aug. 15, 1958; 36 km east of Min'tsin town, somewhat sand-covered, Aug. 18-1959, Petr.), *Ordos* (right bank of Huang He river, opposite Hekou, Aug. 7, 1884 – Pot.; "Bayannur ajmaq [Urot]" – Fl. Intramong. (1993) l.c.).

General distribution: China (Dunbei).

14. A. brachyloba Franch. in Pl. David. 1 (1884) 171; Forbes et Hemsl. Index Fl. Sin. 1 (1888) 442; Kitag. Lin. Fl. Mansh. (1939) 424; Fl. Intramong. 6 (1982) 145; Fl. Sin. 76, 2 (1991) 74; Fl. Intramong. ed. 2, 4 (1993) 619; Gub. Konsp. Fl. Vneshn. Mong. [Conspectus of Flora of Outer Mongolia] (1996) 96. —Ic.: Fl. Intramong. (1982) tab. 51, fig. 1–5; Fl. Intramong. (1993) tab. 243, fig. 1–5.

Described from China (vicinity of Beijing). Type in Paris (P).

On steppe slopes of low mountains, in rock crevices, fringes of solonchaks.

IA. Mongolia: *Cent. Khalkha* (Ikhe-Tukhum-nor lake, Dzhargalant river basin, Aug. 10, 1925 – Krasch.; 5 km eastward of Choiren-ul, steppe on fringe of solonchak lowland, Aug. 28, 1940 – Yun.; steppe slopes of conical hillocks on road to Dalan-Dzadgad from Ulan-Bator, July 11, 1950 – Kal.), *East. Mong.* (southward of Barun-Chulut-Daba, 1899 – Pal.; Khalkha-gol somon, bank of Kerulen river, Aug. 24, 1906 – Novitskii; 185 km southeast of Ulan-Bator, steppe, Aug. 21, 1940 – Yun.; Choibalsan, Bayan-ula, steppe, Aug. 14, 1956; north-east of Enger-Shanda, on rocks, Oct. 11–1959, Dashnyam; "Yakeshi, Shilin-Khoto, Ulantsab (Sinkhe)" – Fl. Intramong. (1993) l.c.); *Ordos* ("Ordos"-Fl. Intramong. (1993) l.c.).

General distribution: Nor. Mong. (Mong.-Daur.), China (Dunbei, East.).

15. A. caespitosa Ledeb. Fl. Alt. 4 (1833) 80; Krasch. in Kryl. Fl. Zap. Sib. 11 (1949) 2801; Grub. Konsp. fl. MNR [Conspectus of Flora of Mongolian People's Republic] (1955) 250; Poljak. in Fl. SSSR, 26 (1961) 499; Fl. Intramong. 8 (1985) 325, excl. syn. *A. frigidioides* H.C. Fu et Z.Y. Zhu; Fl. Sin. 76, 2 (1991) 72; Fl. Intramong. ed. 2, 4 (1993) 616. —Ic.: Ledeb. Ic. pl. fl. ross. 5 (1844) tab. 472.

71 Described from Siberia (Altay). Type in St.-Petersburg (LE).

On barren rubbly and stony slopes of low mountains, outcrops of red Gobi sandstones, solonchaks, solonetzes, on pebble beds of gorges, along banks of rivers and lakes.

IA. Mongolia: *Khobd., Mong. Alt., Cent. Khalkha, East. Mong., Depr. Lakes, Val. Lakes, Gobi Alt., East. Gobi, West. Gobi, Alash. Gobi, Ordos.*

IC. Qaidam: *Plain* (14 km northward of Tsagan-Us, somewhat sand-covered, Oct. 14, 1959 – Petr.).

IIA. Junggar: *Jung. Gobi* (valley of Uinchin-gol river, 4 km before Putsgein-gol estuary, shrubby steppe, Aug. 15, 1979 – Grub.).

IIIA. Qinghai: *Nanshan* (nor. slope) Tutung mountain range, steppe, Aug. 7, 1889 – Przew.; 15 km southward of Aksai settlement, rocky slopes of Altyntag mountain range, Aug. 2, 1958; 33 km west of Xining town, rocky slopes of mounds, Aug. 5; Kukunor lake, on bank, Aug. 5 – 1959, Petr.).

General distribution: West. Sib. (Altay), Nor. Mong. (Hang.).

16. A. campbellii Hook. f. et Thoms. ex Clarke, Comp. Ind. (1876) 164; Hemsl. Fl. Tibet (1902) 182; Pamp. Fl. Carac. (1930) 212; Enum. vasc. pl. Xizang. (1980) 361; Fl. Xizang. 4 (1985) 763; Fl. Sin. 76, 2 (1991) 134.

Described from Himalayas (Kashmir). Type in London (K).

In mountain steppes, valleys of mountain rivers, lake depressions, rarely in talus.

IIIB. Tibet: *Weitzan* (Russkoe lake (Orin-nor), July 27, 1884 — Przew.; "Chantan, Shuankhu, Tuzyantszy, Chzhunba, Chzhada, Shigapzy" — Fl. Xizang. (1985) l.c.).

General distribution: Himalayas (Kashmir).

Note. Steppe plants of mountain regions; in laciniated leaf blade, close to the group of species *A. santolinifolia* (Turcz. ex Pamp.) Krasch. and *A. gmelinii* Web. ex Stechm. and, in live form (not subshrubs), to the group of montane mesoxerophytes of type *A. vexans* Pamp. et al in pubescence, form of anthodia and their arrangement in the panicle.

17. A. chamomilla Winkl. in Acta Horti Petrop. 10 (1887) 87.

Described from Sinkiang (Jung. Alat.). Type in St.-Petersburg (LE).

IIA. Junggar: *Jung. Alat.* (Borotala river, 1500–2000 m above sea, Aug. 1878 — A. Reg., typus!; Borotala river, Aug. 1879 — A. Reg.), *Tien Shan* (from Kul'dzha and Dzhagastai, at crossings, Aug. 21, 1957 — Yun. et Yuan').

General distribution: endemic.

18. A. dalai-lamae Krasch. in Not. syst. Herb. Horti Petrop. 3 (1922) 17; Enum. vasc. pl. Xizang. (1980) 362; Fl. Intramong. 6 (1982) 143; Fl. Xizang. 4 (1985) 759; Fl. Sin. 76, 2 (1991) 69; Fl. Intramong. ed. 2, 4 (1993) 621. — Ic.: Fl. Intramong. (1982) tab. 51, fig. 6.

Described from Tibet. Type in St.-Petersburg (LE). Plate VI, fig. 2. Map 6.

In mountain barren steppes, in mountains up to 3000 m and above.

IA. Mongolia: *Alash. Gobi* (Bayan-khoto, Tengri, Dzhalamu, Aug. 15, 1958 — Petr.; "Toushan'" — Fl. Intramong. (1993) l.c.), *Khesi* (95 km east — south-east of Chzhan'e and 80 km west — north-west of Yunchan on road to Lanzhou from An'si, intermontane valley north of Richthofen mountain range, barren steppe, Oct. 8, 1957 — Yun.; 45 km west of Yunchan town, hilly Nanshan foothills, July 10, 1958 — Petr.).

IC. Qaidam: *Plain* (Tsagan-Us, 15 km from Chzhak, on nor. bank of Chzhak lake, Oct. 15, 1959 — Petr.), *Mount.* (Qaidam, Nanshan, Aug. 1879 — Przew.; paratypus!; Dulan-khit temple, 1901 — Lad.).

IIIA. Qinghai: *Nanshan* (Machan-Shan' mountain range, on loam, July 16, 1908 — Czet., paratypus!; 20 km west of Gunkho town, Aug. 6, 1959 — Petr.).

IIIB. Tibet: *Weitzan* (Burkhan-Budda mountain range, Nomokhungol, Aug. 3–15, 1884 — Przew., typus!; same site, nor. slope of Khatu gorge, July 27, 1901 — Lad., paratypus!).

General distribution: China (Nor.).

19. A. disjuncta Krasch. in Not. syst. (Leningrad) 9 (1946) 176; Grub. Konsp. fl. MNR [Conspectus of Flora of Mongolian People's Republic] (1955) 264; Leonova in Grub. Opred. rast. Mong. [Key to Plants of Mongolia] (1982) 250; Fl. Sin. 76, 2 (1991) 24; Gub. Konsp. fl. Vneshn. Mong. [Conspectus of Flora of Outer Mongolia] (1996) 97.

Described from Sinkiang (Tien Shan). Type in St.-Petersburg (LE).

On rocky and stony slopes, among rocks in alpine and subalpine belts of mountains, 3100–4000 m above sea.

IA. Mongolia: *Mong. Alt.* (Adzhi-Bogdo mountain range, Ikhe-gol river, sheep's fescue-*Cobresia* slopes, July 22, 1979—Grub., Muld., Dar.), *Gobi Alt.* (Ikhe-Bogdo mountain range, upper Baga-Artsatuin creek valley, among rocks, Sept. 8, 1943—Yun.; same site, Nariin-Khurimt gorge, on rocks, July 28, 1948—Grub.).

IIA. Junggar: *Tien Shan* (Biangou town, Sept. 25, 1929—Pop., typus!; Manas river basin, valley of Danu-gol river, in rock crevices, July 2, 1959; left bank of Ulan-usu river at confluence with Dzhartas river, rocky slopes, July 18, 1957—Yun., Yuan' et Li).

General distribution: endemic.

20. A. feddei Levl. et Vaniot, Feddes Repert. 8 (1910) 138; Gub. Konsp. fl. Vneshn. Mong. [Conspectus of Flora of Outer Mongolia] (1996) 197.— *A. feddei* subsp. *arschantinica* (Darijmaa) Gubanov et R. Kam. in Gub. l.c. 197.— *A. lavandulifolia* DC. Prodr. 6 (1838) 110, non Salisb. 1796; Poljak. in Fl. SSSR, 26 (1961) 453; Fl. Intramong. 6 (1982) 141.— *A. lancea* Van. in Bull. Acad. Internat. Giorn. 12 (1903) 500; Fl. Intramong. ed. 2, 4 (1993) 627.— *A. arschantinica* Darijmaa in Bull. Soc. natur. Moscou, 97, 5 (1992) 65.

Described from China. Type in Geneva (G).

In forest grasslands, meadows among shrubs, in willow thickets, along river banks, sometimes as weed.

IA. Mongolia: *East. Mong.* (vicinity of Khailar town, bank of Argun' river, July 18, 1909—Litw.; Choibalsan, 15 km north-west of Bayanty-bulak, willow thickets, Aug. 22, 1949—Yun.; vicinity of Khailar town, sandy site on river bank, Aug. 29, 1951—A.R. Lee Buyar-nur lake, marshy meadow, March 18, 1977—Dashnyam; "Shilin-Khoto, Khukh-Khoto"—Fl. Intramong. (1993) l.c.), *Alash. Gobi* (Arshantyn-nuru mountains, 8 km from Shulin post, near spring, Aug. 1, 1998, No. 235—Grub., Gub., Dar.).

General distribution: Far East, Nor. Mong. (Fore Hing.), China (Nor., East.), Korean peninsula.

Note. The find of species *A. princeps* Pamp. closely related to *A. feddei* Levl. et Vaniot, is possible in East. Mongolian territory as reported in Fl. Intramong. (1993) for Shilin-Khoto town.

21. A. freyniana (Pamp.) Krasch. in Spis. rast. Gerb. fl. SSSR [Catalogue of Plants in the Herbarium of Flora of USSR] 9 (1949) 42, No. 3277; Grub. Konsp. fl. MNR [Conspectus of Flora of Mongolian People's Republic] (1955) 264; Poljak. in Fl. SSSR, 26 (1961) 467; Leonova in Grub. Opred. rast. Mong. [Key to Plants of Mongolia] (1982) 247; Fl. Intramong. 8 (1985) 322; Fl. Sin. 76, 2 (1991) 52; Fl. Intramong. ed. 2, 4 (1993) 614; Gub. Konsp. fl. Vneshn. Mong. [Conspectus of Flora of Outer Mongolia] 73 (1996) 97.— *A. sacrorum* Ledeb. var. *latifolia* Ledeb. f. *freyniana* Pamp. in Nouv. Giorn. Bot. Ital. n.s. 34 (1927) 688. —Ic.: Fl. Intramong. (1993) tab. 240, fig. 9, 10.

Described from East. Siberia. Type in Firenze (Florence) (FI).

On steppe rocky slopes, meadows, forest borders, along gorge floors, on pebble beds.

IA. Mongolia: *Khobd., Cent. Khalkha, Depr. Lakes, East. Gobi, Gobi Alt.* (Gub. l.c.), *East. Mong.* (Dzalal-khan somon, nor.-west of extinct volcano, on basalts, July 15, 1944 – Yun.; Dapgabalbar somon, 15 km before Shair-Togoo, on Tucha river bank, Aug. 2, 1963 – Dashnyam; 49 km north of Barun-Urta on road to Choibalsan, steppe, July 14, 1971 – Dashnyam, Karam. et Safronova; "Ulantsab, Inypan', Shilin-Khoto, Khukh-Khoto" – Fl. Intramong. (1993) l.c.), *Alash Gobi* ("East. Alashan [Helanshan'], Baoto [Alashan]" – Fl. Intramong. (1993) l.c.).

General distribution: Far East (south.), Nor. Mong. (Hent., Mong.-Daur.), China (Dunbei).

22. **A. frigida** Willd. Sp. pl. 3, 3 (1803) 1838; Kitag. Lin. Fl. Mansh. (1939) 425; Krasch. in Kryl. Fl. Zap. Sib. 11 (1949) 2797; Grub. Konsp. fl. MNR [Conspectus of Flora of Mongolian People's Republic] (1955) 264; Poljak. in Fl. SSSR, 26 (1961) 494; excl. syn. *A. argyrophylla* Ledeb.; Fl. Kazakhst. 9 (1966) 98; Leonova in Grub. Opred. rast. Mong. [Key to Plants of Mongolia] (1982) 250; Fl. Intramong. 6 (1982) 156; Pl. vasc. Helansch. (1986) 357; Fl. Sin. 76, 2 (1991) 15; Fl. Intramong. ed. 2, 4 (1993) 606; Opred. rast. Sr. Azii [Key to Plants of Mid. Asia] 10 (1993) 555; Gub. Konsp. fl. Vneshn. Mong. [Conspectus of Flora of Outer Mongolia] (1996) 97. – *A. frigida* var. *typica* and var. *intermedia* Trautv. in Bull. Soc. natur. Moscou, 2 (1866) 358. – *A. frigida* subsp. *willdenowiana* (Bess.) Krasch. in Krasch. l.c. 2798. – *A. frigida* subsp. *gmeliniana* (Bess.) Krasch. et Kryl. l.c. 2798. – *A. frigida* Willd. subsp. *parva* Krasch. in Krasch. l.c. 2798. – *A. absinthium* var. *gmeliniana* and var. *willdenowiana* Bess. in Syn. Absinth. (1829) 251-254. – *A. davazamczii* Darijmaa et R. Kam. in Bull. Soc. natur. Mosc. 97, 5 (1992) 65; Gub. Konsp. Fl. Vneshn. Mong. [Conspectus of Flora of Outer Mongolia] (1996) 97. – Ic.: Fl. Intramong. (1982) tab. 45, fig. 1-6; Fl. Intramong. (1993) tab. 239, fig. 1-6.

Described from Siberia. Type in Berlin (B).

In arid barren and sandy steppes, on rubble trails of mountains and conical hillocks, sometimes on sand mounds and pebble beds of rivers and gorges, along fringes of pine forests, rarely on fallow land, in chee grass thickets, common and massive. Edificator (characteristic species).

IA. Mongolia: *Khobd., Mong. Alt., Cent. Khalkha, East. Mong., Depr. Lakes, Gobi Alt., East. Gobi, West. Gobi, Alash. Gobi, Ordos.*

IIA. Junggar: *Tarb.* (Saur mountain range on road to Burchim from Kosh-Tologoi, steppe, July 4; on road to Altay from Karamai, Aug. 4–1959, Yun.), *Jung. Alt.* (Dzhair mountain range, 25 km west of Aktam settlement on road to Chuguchak, conical hillock slopes, Aug. 3; Maili mountain range, 20 km north-east of meteorological station in Junggar gateway toward Karaganda pass, conical hillocks, Aug. 14–1957, Yun., Lee et Yuan'), *Tien Shan* (Sairam-nur lake, July 20, 1878 – A. Reg.; Bogdo-ula mountain range, rocky slopes of foothills, Sept. 14, 1929 – Pop.; Kuitun river basin, Bain-gol creek valley south of Tushandza town, steppe, June 29, 1957 – Yun.), *Jung. Gobi* (Baga-Khabtak-nuru mountain range, rocky nor. slopes, Sept. 14; Takhilty-ula area, on nor. slopes, Sept. 17–1948, Grub.; 30 km north-west of Bulgan settlement, near Ikh-Dzhargalant post on Chinese border, July 6; east. macroslope of Buitag-Bogdo mountain range, Budun-Kharkaityn-gol river basin, July 28–1979, Gub.), *Zaisan* (Maikapchagai mountains, rocky slopes, June 6; Chern. Irtysh, left bank west of Cherektas mountain, sandy-pebble bed steppe, July 11–1914, Schischk.; Kobuk river valley, barren rubble slopes, July 20, 1914 — Sap.).

General distribution: Fore Balkh., Jung.-Tarb., Europe, West. and East. Sib., Far East, Nor. Mong., Nor. Amer.

74   23. A. glabella Kar. et Kir. in Bull. Soc. natur. Moscou, 14 (1841) 44; Krasch. in Kryl. Fl. Zap. Sib. 11 (1949) 2793; Fl. Kazakhst. 9 (1966) 97; Fl. Sin. 76, 2 (1991) 74; Opred. rast. Sr. Azii [Key to Plants of Mid. Asia] 10 (1993) 554; Gub. Konsp. fl. Vneshn. Mong. [Conspectus of Flora of Outer Mongolia] (1996) 98. — *A. obtusiloba* auct. non Ledeb.: Poljak. in Fl. SSSR, 26 (1961) 510. — Ic.: Fl. Kazakhst. Plate 11, fig. 2.

Described from East. Kazakhstan (Bukhtarma). Type in St.-Petersburg (LE).

On rocky slopes of low mountains and conical hillock, along arid pebble beds, among rocks.

IA. Mongolia: *Khobd.* (midcourse of Tsagan-Nuriin-gol, Obgorula mountain, lower belt of mountains, Aug. 2, 1945 — Yun.), *Depr. Lakes* (3 km south of Ulangom on road to Kobdo, granitic conical hillock, rocky slope, July 28, 1945 — Yun.).

General distribution: Fore Balkh., Jung.-Tarb., West. Sib. (Altay), Mong. (Hang.).

Note. Closely related to *A. obtusiloba* Ledeb. differing from it in glabrous thalamus, low fertile shoots and punctate-glandular-pubescent leaf blade.

24. A. gmelinii Web. ex Stechm. Artem. (1775) 17; Krasch. in Kryl. Fl. Zap. Sib. 11 (1949) 2790; Grub. Konsp. Fl. MNR [Conspectus of Flora of Mongolian People's Republic] (1955) 265; Poljak. in Fl. SSSR, 26 (1961) 464, excl. *A. messerschmidtiana* Bess.; Fl. Kazakhst. 9 (1966) 91; Leonova in Grub. Opred. rast. Mong. [Key to Plants of Mongolia] (1982) 247; Fl. Intramong. 6 (1982) 152; Pl. vasc. Helansch. (1986) 258; (1987) 634; Fl. Sin. 76, 2 (1991) 47; Fl. Intramong. ed. 2, 4 (1993) 613, excl. syn. *A. santolinifolia* Turcz. ex Bess.; Opred. rast. Sr. Azii [Key to Plants of Mid. Asia] 10 (1993) 551; Gub. Konsp. fl. Vneshn. Mong. [Conspectus of Flora of Outer Mongolia] (1996) 97. — *A. gmelinii* var. *intermedia* Ledeb. Fl. Alt. 4 (1833) 72. — *A. sacrorum* Ledeb. in Mem. Acad. Sci. Petersb. 5 (1812) 71; Franch. Pl. David. 1 (1884) 170, p.p., Forbes et Hemsl., Index Fl. Sin. 1 (1888) 444, p.p.; Hemsl. Fl. Tibet. (1902) 183, p.p.; Pamp. Fl. Carac. (1930) 211, p.p.; Kitag. Lin. Fl. Mansh. (1939) 439, p.p. — *A. iwayomogi* Kitam. in Acta Phytotax. Geobot. 7, 2 (1938) 64. — *A. polybotryoidea* Y.R. Ling in Bull. Bot. Res. 5, 2 (1985) 1. — Ic.: Ledeb. Ic. pl. Ross. 5 (1834) fig. 470; Fl. SSSR, Plate 23, fig. 2; Fl. Kazakhst. Plate 10, fig. 2; Fl. Intramong. (1982) tab. 52,

fig. 1–6; Drevesn. rast. Tsinkhaya [Wooded Plants of Qinghai] (1987) 462; Intramong. (1993) tab. 240, fig. 7, 8.

Described from East. Siberia. Type in St.-Petersburg (LE). Map 5.

On steppe rock slopes of mountains, rocks, among scrubs, on talus.

IA. Mongolia: *Mong. Alt.* (south-west. slope of Mongolian Altay, 30 km north-west of Bulgan settlement, not far from Ikh-Dzhergalant post, Aug. 6, 1979 — Gub.), *Cent. Khalkha* (30 km south of Underkhan, on rocky slopes, July 10, 1971 — Isach.; 180 km south — south-west of Ulan-Bator, on old road to Dalan-Dzadagad, July 15, 1943 — Yun.), *East. Mong.* (Choibalsan, 13–15 km west of Shavorte-nur lake, rocky steppe, July 19, 1949 — Yun.; 30 km south-east of Choibalsan town, steppe, Aug. 28, 1954; 8 km west of Tumen-Delger somon, rocky steppe slopes of conical hillocks, Aug. 7, 1956; 30 km east of Enger-Shand, floodplain of Kerulen river, Aug. 9, 1956 — Dashnyam; "Shilin-Khoto, Khukh-Khoto, Baotou [Datsinshan'], Ulantsab" — Fl. Intramong. (1993) l.c.), *Gobi Alt.* (Dzun-Saikhan mountain range, mountain steppe, Oct. 8, 1940 — Yun.), *East. Gobi* (nor. fringe of Khan-Bogdo hummocky massif, rocky slopes, Sept. 28, 1940 — Yun.), *Alash. Gobi* (Alashan mountain range, Yamato area, midbelt of mountains, among scrubs, July 7, 1908 — Czet.), *Ordos* ("Ordos" — Fl. Intramong. l.c.).

General distribution: Jung.-Tarb., Nor. and Cent. Tien Shan, West. and East. Sib., Far East, North Mong. (Hent., Hang., Mong.-Daur., Fore Hing.), China (Nor. and West.).

25. A. grubovii Filat. in Bot. zhurn. 71, 11 (1986) 1553. — Ic.: l.c. fig. 1.

Described from Ordos. Type in St.-Petersburg (LE). Map 5.

In meadows, river valleys and lake basins.

IA. Mongolia: *Ordos* (up along Huang He river up to crossing on Khurei-Khunda river, Aug. 6, 1871 — Przew., typus!; Bain-nur lake, meadow on bank, Sept. 2, 1884 — Pot., paratypus!).

General distribution: China (Dunbei).

26. A. hedinii Ostenf. in Hedin. S. Tibet, 6, 3 (1922) 41; Enum. vasc. pl. Xizang. (1980) 362; Fl. Intramong. 6 (1982) 146; Fl. Xizang. 4 (1985) 754; Pl. vasc. Helansch. (1986) 257; Fl. Sin. 76, 2 (1991) 65; Fl. Intramong. ed. 2, 4 (1993) 618; Opred. rast. Sr. Azii [Key to Plants of Mid. Asia] 10 (1993) 553. — Ic.: Hedin, l.c. tab. 3, fig. 1; Fl. Intramong. (1982) tab. 52, fig. 7, 8; Fl. Xizang. (1985) tab. 325, fig. 1–4; Fl. Intramong. (1993) tab. 242, fig. 7, 8.

Described from Tibet. Type in Stockholm (S).

In valleys of mountain rivers, on moraines at 3300–4800 m above sea.

IA. Mongolia: *Alash.* ("Alashan', Helan'shan', Luntoushan'" — Fl. Intramong. (1993) l.c.).

IIIA. Qinghai: *Nanshan* (Kuku-nor lake, 1908; road to Kuku-nor lake from Alashan, 1909 — Czet.; Mon'yuan', glacier moraine at head of Gantsig river, right tributary of Peishikhe river, 3900–4300 m above sea, Aug. 18; same site, left tributary of Peishikhe river entering Togungkhe river, 3350–3720 m above sea, Aug. 18–1958, Petr.).

IIIB. Tibet: *Chang Tang* ("Ban'ge, Shigatsze, Gaitsze" — Fl. Xizang. l.c.), *Weitzan* (Mekong river basin, Barchyu river, 3600 m, Oct. 1900; Dzhagyn-gol river, Aug. 14, 1905 — Lad.), *South.* (Lhasa, May 1904 — Walton; "Sosyan', An'do, Nan'mulin" — Fl. Xizang. l.c.).

IIIC. *Pamir* (Chimgan-Bashi, July 10, 1909 — Divn.).

General distribution: East. Pamir, Himalayas (West. Kashmir).

27. A. igniaria Maxim. Mem. Pres. Acad. Sci. Petersb. Div. Sav. 9 (1895) 161 [Prim. Fl. Amur.]; Kitag. Lin Fl. Mansch. (1939) 426; Fl. Intramong. 6 (1982) 132; Fl. Sin. 76, 1 (1991) 123; Fl. Intramong. ed. 2, 4 (1993) 635. — *A. vulgaris* auct. non L.: Forbes at Hemsl. Index Fl. Sin. 1 (1884) 446. — Ic.: Fl. Intramong. (1982) tab. 47, fig. 1–6; Fl. Intramong. (1993) tab. 248, fig. 1–6.

Described from Nor. China (vicinity of Beijing). Type in St.-Petersburg (LE).

Along mountain slopes, waste lands, around residences.

IA. Mongolia: *East. Mong.* ("Shilin-Khoto Dolun' town, Khukh-Khoto town" — Fl. Intramong. (1993) l.c.), Alash. Gobi (Alashan mountain range, road to Tarlyk well, July 16, 1873 — Przew.).

General distribution: China (Nor., East.), Korean peninsula.

Note. Differs distinctly from *A. vulgaris* L. in large, slender, somewhat incised leaf blade and sessile anthodia.

28. A. integrifolia L. Sp. pl. (1753) 848; Kitag. Lin. Fl. Mansch. (1939) 420; Krasch. in Kryl. Fl. Zap. Sib. 11 (1949) 2815; Poljak. in Fl. SSSR, 26 (1961) 449; Grub. Konsp. fl. MNR [Conspectus of Flora of Mongolian People's Republic] (1955) 265; Leonova in Grub. Opred. rast. Mong. [Key to Plants of Mongolia] (1982) 248; Fl. Intramong. 6 (1982) 129; Fl. Sin. 76, 2 (1991) 125; Fl. Intramong. ed. 2, 4 (1993) 629; Gub. Konsp. fl. Vneshn. Mong. [Conspectus of Flora of Outer Mongolia] (1996) 92. — *A. vulgaris* auct. non L.: Forbes et Hemsl. Index Fl. Sin. 1 (1884) 446. — Ic.: Fl. Intramong. (1982) tab. 49, fig. 1–5; Fl. Intramong. (1993) tab. 245, fig. 1–5.

Described from Siberia. Type in London (Linn.).

In larch forests, their borders and glades, coastal meadows, along margin of herbaceous swamps, river banks.

76   IA. Mongolia: *East. Mong.* (Choibalsan, Khuntu somon, meadow on floor of marshy creek valley, Aug. 12, 1949 — Yun.; "Yakeshi [Shilin-Kheto town]" — Fl. Intramong. (1993) l.c.).

General distribution: West. and East. Sib., Far East, Nor. Mong., China (Nor.), Korean peninsula.

29. A. jounghusbandii Drumm. ex Pamp. in Nouv. Giorn. Bot. Ital. n.s. 34 (1927) 798; Enum. vasc. pl. Xizang. (1980) 366; Fl. Xizang. 4 (1985) 751; Fl. Sin. 76, 2 (1991) 28.

Described from Tibet. Type in Florence (FI). Isotype in St.-Petersburg (LE).

IIIB. Tibet: *Chang Tang* ("Dansyun, Gaitsze, Zhitu" — Fl. Xizang. (1985) l.c.), *South.* (Tszyantszy, July–Sept. 1904 — Walton, isotypus!; "Nan'mulin, Lhasa, Sage, Lankatsza" — Fl. Xizang. (1985) l.c.).

General distribution: endemic.

Note. This alpine desert species is closely related to *A. rutifolia* Stechm. ex Spreng.

30. A. junatovii Filat. in Bot. zhurn. 71, 11 (1986) 1552. —Ic.: l.c. fig. 2.

Described from Tibet (Weitzan). Type in St.-Petersburg (LE). Map 5.

On mountain steppe slopes.

IC. Qaidam: *Mount.* (Dulan-khit temple, Aug. 8, 1901—Lad.).

IIIA. Qinghai: *Nanshan* (Kuku-nor lake, Aug. 7, 1880—Przew.; slopes of Yuzhno-Kuku-nor mountain range, Aug. 28, 1894—Rob.; 24 km south of Xining town, cereal grass-forbs mountain steppe, Aug. 4; 30 km east of Gunkho town, willow thickets, Aug. 7-1959, Petr.).

IIIB. Tibet: *Weitzan* (Burkhan-Budda mountain range, nor.—nor.-west of Salgin-gol river toward Urgu-Shirik well, Sept. 25, 1879—Przew., typus!).

General distribution: endemic.

Note. Species related to *A. phaeolepis* Krasch.

31. A. laciniata Willd. Sp. pl. 3, 3 (1803) 1843; Franch. Pl. David. 1 (1882) 171; Forbes et Hemsl. Index Fl. Sin. 1 (1888) 444; Kitag. Lin. Fl. Mansh. (1939) 427; Krasch. in Kryl. Fl. Zap. Sib. 11 (1949) 2806; Grub. Konsp. fl. MNR [Conspectus of Flora of Mongolian People's Republic] (1955) 267; Poljak. in Fl. SSSR, 26 (1961) 473; Fl. Kazakhst. 9 (1966) 88; Leonova in Grub. Opred. rast. Mong. [Key to Plants of Mongolia] (1982) 247; Fl. Intramong. 6 (1982) 148; Opred. rast. Sr. Azii [Key to Plants of Mid. Asia] 10 (1993) 550; Gub. Konsp. fl. Vneshn. Mong. [Conspectus of Flora of Outer Mongolia] (1996) 97.—*A. tanacetifolia* auct. non L.: Fl. Intramong. ed. 2, 4 (1993) 614. —Ic.: Grub. Opred. rast. Mong. [Key to Plants of Mongolia] (1982) Plate 134, fig. 615; Fl. Intramong. (1982) tab. 53. fig. 1-6.

Described from Siberia. Type in St.-Petersburg (LE).

In larch, aspen and birch forests and their borders, flood-plain and forest meadows, meadow and steppe slopes of mountains, among shrubs.

IA. Mongolia: *Mong. Alt.* (30 km north-west of Bulgan somon centre, Aug. 6; in Saksai river valley, Aug. 9-1979, Gub.), *East. Mong.* (Choibalsan, 17 km north-east of Kharanur lake, wild rye meadow, Aug. 18, 1949—Yun.; 80 km south-east of Khamar-Daba settlement on Khalkhin-gol river, forbs steppe, July 28, 1971—Karam.; 90 km east of Sain-Shavda town, in gorge, Aug. 5, 1982—Gub.), *Depr. Lakes* (Ulangom, Sept. 6, 1879—Pot.; Khirgiz-nur lake, 11 km from Khirgis somon centre, forbs-oat grass steppe, July 15, 1973—Banzragch; 10 km east of Ulangom town, desertified knolls, Aug. 25, 1979—Gub.).

IIA. Junggar: *Jung. Alt.* (Borotala river basin, south. slope of Junggar Ala Tau, July 21, 1909—Lipsky), *Jung. Gobi* (Baitag-Bogdo mountain range, valley of Budun-Khariityn-gol river, July 29, 1979—Gub.).

General distribution: Jung.-Tarb., Europe, West. and East. Sib., Far East, Nor. Mong. (Fore Hubs., Hent., Hang., Mong.-Daur.), China (Dunbei).

32. A. latifolia Ledeb. in Mem. Acad. Sci. Petersb. 5 (1815) 569; Forbes et Hemsl. Index Fl. Sin. 1 (1888) 444; Kitag. Lin. Fl. Mansh. (1939) 427; Krasch. in Kryl. Fl. Zap. Sib. 11 (1949) 2810; Grub. Konsp. fl. MNR [Conspectus of Flora of Mongolian People's Republic] (1955) 266; Poljak. in Fl. SSSR, 26 (1961) 473; Fl. Kazakhst. 9 (1966) 88; Leonova in Grub. Opred. rast. Mong. [Key to Plants of Mongolia] (1982) 247; Fl. Intramong. 6 (1982) 148; Fl. Sin. 76, 2 (1991) 53; Fl. Intramong. ed. 2, 4 (1993) 608; Opred. rast. Sr. Azii [Key to Plants of Mid. Asia] 10 (1993) 550; Gub. Konsp. fl. Vneshn. Mong. [Conspectus of Flora of Outer Mongolia] (1996) 98. — Ic.: Fl. Kazakhst. Plate 10, fig. 1; Fl. Intramong. (1982) tab. 53, fig. 7, 8; Fl. Intramong. (1993) tab. 241, fig. 7, 8.

Described from Siberia. Type in St.-Petersburg (LE).

On steppe slopes, floodplain, forest and steppe meadows, borders of larch forests, in willow thickets, on pebble beds, rarely on solonetzic meadows.

IA. Mongolia: *East. Mong.* (Shavorte-obo area, ravine with solonetzic soils, Aug. 20, 1949 — Yun.; 10 km from Choibalsan town, in Kerulen river floodplain, Aug. 20, 1957 — Dashnyam; Khalkhin-gol river, meadow with willow thickets, Aug. 18, 1977 — Dar.; "Yakeshi" — Fl. Intramong. (1993) l.c.). "Ordos" — Fl. Intramong. (1993) l.c.).

General distribution: Jung.-Tarb., Europe, West. and East. Sib., Far East, Nor. Mong. (Hent., Mong.-Daur., Fore Hing.).

33. A. leucophylla (Turcz. ex Bess.) Pamp. in Nouv. Giorn. Bot. Ital. n.s. 36 (1930) 414; Krasch. in Kryl. Fl. Zap. Sib. 11 (1949) 2814; Grub. Konsp. fl. MNR [Conspectus of Flora of Mongolian People's Republic] (1955) 266; Poljak. in Fl. SSSR, 26 (1961) 443; Enum. vasc. pl. Xizang. (1980) 363; Leonova in Grub. Opred. rast. Mong. [Key to Plants of Mongolia] (1982) 249; Fl. Intramong. 8 (1985) 323; Fl. Sin. 76, 2 (1991) 105; Fl. Intramong. ed. 2, 4 (1993) 630; Gub. Konsp. fl. Vneshn. Mong. [Conspectus of Flora of Outer Mongolia] (1996) 98. — *A. vulgaris* L. var. *leucophylla* Turcz. ex Bess. in Nouv. Mem. Soc. natur. Moscou, 3 (1834) 54; Clarke, Comp. Ind. (1875) 162, in nota. — *A. mongolica* auct.: N. Wang et H.T. Ho, Fl. Intramong. 6 (1982) 140. — Ic.: Fl. Intramong. (1985) tab. 142, fig. 6, 7; Fl. Intramong. (1993) tab. 247, fig. 6, 7.

Described from Mongolia (Hubsugul lake). Type in St.-Petersburg (LE).

On steppe slopes, coastal solonetzic meadows, in willow thickets, on floors of creek valleys and river valleys.

IA. Mongolia: *Khobd.* (floodplain of Buku-Muren river, cereal grass meadow, July 24, 1977 — Karam.), *Mong. Alt.* (Khobdo river, Aug. 25, 1879 — Pot.), *East. Mong.* ("Khailar, Shilin-Khoto, Khukh-Khoto" — Fl. Intramong. (1993) l.c.), *Depr. Lakes* (at crossing on Kobdo-Ulangom river, Aug. 23, 1944 — Yun.), *Gobi Alt.* (Bain-Gobi somon, Tsagan-gol river, in derris grove, July 27, 1948 — Grub.; Dundu-Saikhan mountain range,

Elyn-ama gorge, July 22, 1970—Grub.), *Alash. Gobi* ("Helanshan'"—Fl. Intramong. (1993) l.c.), *Khesi* (between Nanshan and Alashan mountain ranges, Aug. 17, 1880—Przew.).

IIIB. Tibet: Weitzan (Burkhan-Budda mountain range, on Khatu-gol river, March 18, 1884—Przew.).

General distribution: West. and East. Sib., Nor. Mong. (Hent., Hang., Mong.-Daur.), China (Nor., East.).

Note. *A. leucophylla* as a species name was first proposed by Turczaninov (based on herbarium specimen from Hubsugul lake (Mongolia) preserved in the herbarium of the Komarov Botanical Institute at St.-Petersburg). In 1834, V. Besser used this name and regarded the herbarium specimen from Hubsugul lake as a variety of *A. vulgaris* L. and gave Latin diagnosis. S.B. Clarke in the work "Compositae Indicae" referring to the manuscript of T. Thomson, cited *A. leucophylla* also as a variety of *A. vulgaris* L. but without Latin diagnosis. Pampanini for the first time assigned *A. leucophylla* to the rank of species citing as type (Hubsugul lake) and other herbarium specimens pertaining to this taxon.

78   34. A. macrantha Ledeb. in Mem. Acad. Sci. Petersb. 5 (1815) 573; Bess. in Nouv. Mem. Soc. natur. Moscou, 3 (1834) 34; Ledeb. Fl. Alt. 4 (1833) 76; DC. Prodr. 6 (1838) 109; Krasch. in Kryl. Fl. Zap. Sib. 11 (1949) 2805; Poljak. in Fl. SSSR, 26 (1961) 462; Fl. Kazakhst. 9 (1966) 90; Fl. Sin. 76, 2 (1991) 41; Opred. rast. Sr. Azii [Key to Plants of Mid. Asia] 10 (1993) 552; Gub. Konsp. fl. Vneshn. Mong. [Conspectus of Flora of Outer Mongolia] (1996) 98.

Described from Siberia. Type in St.-Petersburg (LE).

On rocky steppe slopes, in scrubs.

IA. Mongolia: *Mong. Alt.* (Gub. l.c.).

General distribution: Volga-Kam., Trans-Volga, West. and East. Sib., Nor. Mong. (Hent., Hang.).

35. A. macrocephala Jacq. ex Bess. in Bull. Soc. natur. Moscou, 9 (1836) 28; Jacq. Catal. (1833) No. 2000, nom. nud.; Hemsl. Fl. Tibet (1902) 182; Pamp. Fl. Carac. (1930) 213; Krasch. in Kryl. Fl. Zap. Sib. 11 (1949) 2820; Grub. Konsp. fl. MNR [Conspectus of Flora of Mongolian People's Republic] (1955) 266; Poljak. in Fl. SSSR, 26 (1961) 518; Ikonn. Opred. fl. Pamira [Key to Flora of Pamir] (1963) 237; Fl. Kirgiz. 11 (1965) 175; Fl. Kazakhst. 9 (1966) 104; Enum. pl. Xizang. (1980) 363; Leonova in Grub. Opred. rast. Mong. [Key to Plants of Mongolia] (1982) 246; Fl. Xizang. 4 (1985) 747; Fl. Sin. 72, 2 (1991) 7; Opred, rast. Sr. Azii [Key to Plants of Mid. Asia] 10 (1993) 558; Gub. Konsp. fl. Vneshn. Mong. [Conspectus of Flora of Outer Mongolia] (1996) 98. —Ic.: Fl. Xizang. (1985) tab. 320.

Described from Himalayas. Type in Paris (P).

On rubbly, rocky and clayey slopes of mountains, along banks of mountain rivers, on pebble beds, in arid and barren steppes up to high mountains.

IA. Mongolia: *Khobd.* (14 km south of Achit-nur lake, intermontane valley, July 15; Bukhu-Muren river floodplain, sparse pea shrub thickets, July 19–1978, Karam et al.), *Mong. Alt.* (bank of Dzhargalante river, July 20, 1898 – Klem.; Khara-Adzarga mountain range, rocky bank of Naishuren-gol, Aug. 2; Khalyun-gol river, fallow land, Aug. 16; Khasagtu-Khairkhan mountain range, Dundu-Tseren-gol river, rocky trail, Sept. 15–1930, Lob.; Dayan-nur lake, on nor. bank, Aug. 9, 1972 – Metel'tseva; Tal-nur lake, solonchak, July 26, 1980 – Karam. et Sumerina), *Cent. Khalkha* (mid. Kerulen, steppe below Tsetsen-Khan, 1899 – Pal.; Uber-Dzhargalante river basin, meadow, Sept. 4; vicinity of Ikhe-Tukhum lake, terrace, Sept. 6, 1925; from Choiren somon to Narat area, Sept. 8–1927, Krasch. et Zam.; 6 km east of Choiren, arid steppe, Aug. 9; floodplain of Kerulen river, 20 km before Underkhan town and 10 km beyond Muren-gol river estuary, pebble bed, Aug. 17–1970, Mirkin; Undzhul somon, Khariin-nur lake, 40 km east of lake, in chee grass thicket, Aug. 15; same site, north of spring section, Aug. 28–1974, Dar.), *Depr. Lakes* (Buyantu river, plantations, Aug. 28, 1930 – Bar.; nor. bank of Khirgiz-nur lake, old fallow land, Aug. 21, 1944 – Yun.), *Val. Lakes* (on left bank of Tuingol river, on sand, July 12; nor.-east of Ongiin-gol river, on fine rubble, July 29, 1893 – Klem.; near Tatsin-gol river, steppe, July 24, 1924 – Gorbunova; 5–6 km left of Tatsiin-gol river along road to Bayan-Khongor from Ulan-Bator, on slopes, July 29, 1952 – Davazhamts), *Gobi Alt.* (Tostu mountains, Aug. 18; valley of Leg river, Aug. 29–1886, Pot.; Bain-Boro mountains, on rocky slope, Sept. 10, 1931 – Ik.-Gal.; south. slopes of Ikhe-Bogdo mountain range, Narim-Khurimt gorge, July 30; Nemegetu-nura mountain range, on main peak, rocky cone on nor. slope, Aug. 8; Ikhe-Bogdo mountain range, steppe, Aug. 30–1948, Grub.; Barun-Saikhan mountain range, peak, Aug. 23, 1953 – Dashnyam; Tostu-ula mountain range, steppe, Aug. 4, 1976 – Rachk.), *West. Gobi* (Atas-Bogdo mountain range, floor of gorge, Aug. 21, 1976 – Rachk.; Tsagan-Bogdo mountain range, under main peak near spring, in rock crevices, Aug. 29, 1979 – Grub., Muld., Dar.), *Alash. Gobi* (Urdzhyum river valley, southward of Tostu mountain range, Aug. 16, 1886 – Pot.).

IB. Kashgar: *Nor.* (Egin', near ruins of Chinese fort, July 11, 1913 – Knorr.), *West.* (Moin' river, midcourse, Aug. 23, 1913 – Schischk.), *South.* (Keriya, Kyuk-Egil' river, June 28, 1885 – Przew.), *East.* (Algoi river, Sept. 11, 1879 – A. Reg.).

IIA. Junggar: *Tien Shah* (Khaidyk river, seiren-Tokhoi area, Aug. 6, 1893 – Rob.; Santash pass, 1879 – Lar.).

IIIA. Qinghai: *Nanshan* (Nanshan, alpine belt, 1879 – Przew.; Chansai, nor.-west. slope of Humboldt mountain range, July 23, 1895 – Rob.).

IIIB. Tibet: *Weitzan* (Burkhan-Budda mountain range, Aug. 6, 1884 – Przew.; "Zhitu, Gaidze, Shuankhu" – Fl. Xizang. l.c.), South. ("Chzhunba" – Fl. Xizang. l.c.).

IIIC. Pamir (Muztag-Ata foothills, on talus, July 20; Tagarma river valley, along valley border, July 25; Kok-Muinak pass at exit from Tagarma valley, on rocky slopes, July 27–1909, Divn.; Karai village, on bank of Tokesekrik river, July 8, 1913 – Knorr.).

General distribution: Nor. and Cent. Tien Shah, East. Pamir, West. Sib. (Altay), Nor. Mong. (Hing.), China (Nor., Nor.-West., Cent., South-West.), Himalayas (West., Kashmir).

36. A. maximovicziana Krasch. ex Poljak. in Not. Syst. (Leningrad) 17 (1955) 403; Krasch. in Mat. po istor. fl. i rast. SSSR, 2 (1946) 137, nom. nud.; Poljak. in Fl. SSSR, 26 (1961) 469; Fl. Sin. 76, 2 (1991) 53; Fl. Intramong. ed. 2, 4 (1993) 609; Gub. Konsp. fl. Vneshn. Mong. [Conspectus of Flora of Outer Mongolia] (1996) 98. – Ic.: Fl. Sin. 76, 2 (1991) 53; Krasch. l.c. (1946) 120.

Described from Far East. Type in St.-Petersburg (LE).

On meadows, northern slopes of conical hillocks, in sparse deciduous forests, along river valleys.

IA. Mongolia: *East. Mong.* (Vicinity of Khailar town, Aug. 21, 1925—Kozlov); Gurvan-Zagal somon, 15 km nor.-west of Khamar-Daba, Aug. 11, 1948—Yun.; Buir-nur lake, Aug. 18, 1977—Dar.; "Yakeshi, Manchuria town"—Fl. Intramong. (1993) l.c.).

General distribution: Far East, Nor. Mong. (Mong.-Daur.), China (Dunbei).

37. **A. messerschmidtiana** Bess. in Nouv. Mem. Soc. natur. Moscou, 3 (1834) 27; Leonova in Grub. Opred. rast. Mong. [Key to Plants of Mongolia] (1982) 247; Gub. Konsp. fl. Vneshn. Mong. [Conspectus of Flora of Outer Mongolia] (1996) 98.—*A. gmelinii* Web. ex Stechm. var. *messerschmidtiana* (Bess.) Poljak. in Fl. SSSR, 26 (1961) 464.—*A. sacrorum* Ledeb. var. *messerschmidtiana* (Bess.) Y.R. Ling in Bull. Bot. Research, 8, 4 (1988) 13; Fl. Intramong. ed. 2, 4 (1993) 611.

Described from Siberia. Type in St.-Petersburg (LE).

On rocky and rubbly slopes and rocks, coastal meadows.

IA. Mongolia: *Cent. Khalkha* (Undzhul-ula mountains, 30 km from Undzhul somon, July 26; Zhargal-Khairkhan mountains, rocky slope, Aug. 24–1974, Dar; Munkhan-ula mountains, rocky slopes, Aug. 16, 1985—Kam. et Dar.), *East. Mong.* (Cent. Kerulen, Bain-Khan mountains, granitic slopes, 1899 — Pal.; "Shilin-Khoto, Khukh-Khoto, Baotou"—Fl. Intramong. (1993) l.c.), *Gobi Alt.* (Dundu-Saikhan mountains, rubbly slopes, July 13, 1909—Czet.; on Bain-Tsagan mountain slopes, Aug. 4, 1931—Ik.-Gal.; same site, Bain-Dalai, July–Aug. 1933—Khurlat et Simakova), *East. Gobi* (200 km south of Sain-Shanda town, Barun-Nariin-ula mountains, west. offshoot of Khutag-ul mountain range, rubbly slopes, July 8, 1892—Gub.), *Khesi* (25 km south of Kulon, east. fringe of Nanshan mountain range, Aug. 12, 1958—Petr.).

General distribution: East. Sib. (Sayans), Far East (south), Nor. Mong. (Hent., Hang., Mong.-Daur., Fore Hing.), China (Nor. and East.), Korean peninsula.

38. **A. mongolica** (Bess.) Fisch. et Mey. ex Nakai in Bot. Mag. Tokyo, 31 (1917) 112; Grub. Konsp. fl. MNR [Conspectus of Flora of Mongolian People's Republic] (1955) 266; Poljak. in Fl. SSSR, 26 (1961) 444; Leonova in Grub. Opred. rast. Mong. [Key to Plants of Mongolia] (1982) 249; Fl. Intramong. 6 (1982) 138; Fl. Sin. 76, 2 (1991) 111; Fl. Intramong. ed. 2, 4 (1993) 630; Gub. Konsp. fl. Vneshn. Mong. [Conspectus of Flora of Outer Mongolia] (1996) 98.—*A. vulgaris* L. var. *mongolica* Bess. in Nouv. Mem. Soc. natur. Moscou, 3 (1834) 53; Fisch. Catal. Hort. Gorenk. (1812) nom. nud. —Ic.: Fl. Intramong. (1982) tab. 142, fig. 1–5; Fl. Intramong. (1993) tab. 247, fig. 1–5.

Described from Mongolia (Cent. Khalkha). Neotype in St.-Petersburg (LE).

80   In steppes, on slopes of low mountains, in scrubs, on solonetzic meadows.

IA. Mongolia: *Cent. Khalkha* (Ulkhin-Bulak area, Aug. 22, 1925 – Gusev, neotypus!; vicinity of Ikhe-Tukhum-nor lake, June 1926 – Zam.), *East. Mong.* (vicinity of Choibalsan town, in old quarry, Aug. 17, 1957; same site, southward of town, river floodplain, June 3, 1958 – Dashnyam; Dariganga, Shilin-Bogdo-ula, cereal grass-forbs steppe, Aug. 11, 1970 – Grub.), *Depr. Lakes* (westward of Ulangom, July 17, 1930; valley of Barun-Turun river, Sept. 6, 1931 – Bar.), *Val. Lakes* (Saikhan-obo somon, 10 km beyond Khoshu-Khida, valley of Ongiin-gol river, solonchak meadow, July 8, 1941 – Tsatsenkin; same site, 30 km before Khoshu-Khida, July 11, 1948 – Grub.), *Gobi Alt.* (Dundu-Saikhan mountains, near brook, Aug. 20, 1931 – Ik.-Gal.), *East. Gobi* (Khan-Bogdo somon, hummocky massif in Galba, on slopes, Sept. 28, 1940 – Yun.).

IIA. Junggar: *Cis-Alt.* (10 km north – north-west of Ertai settlement, on road to Kosh-Tologoi, arid meadow, July 14, 1959 – Yun.).

General distribution: East. Sib. (Sayans), Nor. Mong. (Fore Hubs., Hent., Hang., Mong.-Daur.).

Note. Type lost. Neotype selected from Gusev's collection from these regions.

39. A. moorcroftiana Wall. ex DC. Prodr. (1838) 117; Wall. Cat. (1828) No. 3296, nom. nud.; Pamp. Fl.. Carac. (1930) 212; Enum. vasc. pl. Xizang. (1982) 363; Fl. Xizang. 4 (1985) 759; Fl. Sin. 76, 2 (1991) 131. – Ic.: Fl. Xizang. (1985) tab. 329.

Described from Himalayas (Kashmir). Type in Geneva (G).

In glades in spruce forests.

IC. Qaidam: *Mount.* (Dulan-Khit temple, 3300 m, Aug. 8, 1901 – Lad.).

IIIB. Tibet: *South.* ("Nan'mulin, Shigapze, Chzhumba, Pulan'), *Chang Tang* (Ben'ge) – Fl. Xizang. (1985) l.c.).

General distribution: China (South-West.), Himalayas (West. Kashmir).

Note. Mountainous mesoxerophilous species related to *A. roxburgiana* Wall. ex Bess. and *A. leucophylla* (Turcz. ex Bess.) Pamp.

40. A. obscura Pamp. in Nouv. Giorn. Bot. Ital. n.s. 36 (1930) 417; Grub. Konsp. Fl. MNR [Conspectus of Flora of Mongolian People's Republic] (1955) 267; Leonova in Grub. Opred. rast. Mong. [Key to Plants of Mongolia] (1982) 249; Gub. Konsp. Fl. Vneshn. Mong. [Conspectus of Flora of Outer Mongolia] (1996) 198.

Described from Mangolia. Syntypes in St.-Petersburg (LE). Map 5.

Along banks and arid beds of rivers, gorge floors, rock trails, as weed along irrigation canals, in pasture lands.

IA. Mongolia: *Mong. Alt.* (Khara-Adzarga mountain range, Nai-turen-gol river, among rocks, Sept. 2; bank of Dundu-Seren-gol river, Khasaktu-Khairkhan mountains, on sand bed, July 13, 1909 – Czet., syntypus!), *Gobi Alt.* (Barun-Saikhan mountain, Sept. 20, 1931 – Ik.-Gal.; Baga-Bogdo mountain range, valley of Tsagan-Burgan river, July 24, 1978 – Ogureeva), *Depr. Lakes* (Chon-Kharykha river, on bank, Aug. 13, 1879 – Pot., syntypus!), *East. Gobi* (Dalan-Dzadagad, weed in irrigated sections, July 21, 1943 – Yun.).

IB. Kashgar: *Nor.* (foot of south. Tien Shan, at Sairas village, near Aksu town, in farms, Aug. 8, 1929 – Pop.), *South.* (Keriya, Aug. 1885 – Przew., syntypus!)

IIA. Junggar: *Tien Shan* (Takes river, July 1, 1893 – Rob., syntypus!), *Jung. Gobi* (Baitag-Bogdo-nuru mountain range, Takhil-tu-Ula, 5 km from Ulyastu-gol estuary, on bank, July 17, 1948 – Grub.).

IIIA. Qinghai: *Nanshan* (vicinity of Pinfan'syan village, July 13, 1875 — Pias.; Tetung mountain range, July 26, lectotypus!; Yuzhno-Tetungsk mountain range, Chertynton temple, Aug. 1–1880, Przew.).

IIIC. Pamir (Tagarma river valley, July 23, 1913 — Knorr.; Issyk-su river estuary, Aug. 19, 1942 — Serp.).

General distribution: China (Nor., West.), Nor. Mong. (Fore Hubs., Hang.).

81   41. A. obtusiloba Ledeb. Fl. Alt. 4 (1833) 68; Krasch. in Kryl. Fl. Zap. Sib. 11 (1949) 2792; Grub. Konsp. fl. MNR [Conspectus of Flora of Mongolian People's Republic] (1955) 277; Poljak. in Fl. SSSR, 26 (1961) 509, excl. syn. *A. glabella* Kar. et Kir.; Fl. Kazakhst. 9 (1966) 96; Leonova in Grub. Opred. rast. Mong. [Key to Plants of Mongolia] (1982) 251; Fl. Sin. 76, 2 (1991) 122; Opred. rast. Sr. Azii [Key to Plants of Mid. Asia] 10 (1993) 554; Gub. Konsp. fl. Vneshn. Mong. [Conspectus of Flora of Outer Mongolia] (1996) 98. — Ic.: Fl. Kazakhst. 9, Plate 11, fig. 1.

Described from Siberia (Altay). Type in St.-Petersburg (LE).

On rocky slopes of mountains and knolls, rocks, in scrubs, on flanks of ravines.

IA. Mongolia: *Khobd.* (near Ureg-nur lake, southward of Takhir-Tologoi mountain, rock outcrops, July 17; Achit-nur lake basin, Tevkh-ula town, nor. bank of Baga-nur lake, on rocks, July 22, 1977, Karam. et Sanczir), *Mong. Alt.* (upper courses of Uinchi river, desert steppe, Sept. 12, 1930 — Bar.; Kukhengir mountains, rocky slopes, Sept. 17, 1930 — Pob.; Delyun somon, east. rocky macroslope, Aug. 6, 1979 — Gub.), *Depr. Lakes* (25 km westward of Tsagan-01 on road to Khobdo, rocky steppe, March 28, 1944; 40 km north of Khobdo, rocky slope, Aug. 7; Dzun-Gobi somon, Borig-del' sand, 7 km south-east of Baga-nur lake, July 25–1945, Yun.; 8 km west of Ulangom, on flank of gorge, July 4, 1977 — Karam. et Sanczir), *Gobi Alt.* (Ikhe-Bogdo mountain range, midportion of Nokhoituin-Khundei creek valley, arid steppe, Sept. 14, 1943 — Yun.; Nemegetu mountain range, 35 km north of Gurvan-Tes settlement, July 29, 1972 — Rachk. et Guricheva).

IIA. Junggar: *Tarb.* (Saur mountain range, valley of Karagaitu river at its exit onto trails, rocky flank of valley, June 23, 1957 — Yun., Li et Yuan'), *Jung. Gobi* (25 km south-west of Bulgan somon, Barangiin-Khara-nuru mountain range, in ravine, Aug. 6, 1977 — Rachk. et Volk.).

General distribution: Fore Balkh., Jung.-Tarb., West. Sib. (Altay), Nor. Mong. (Hang.).

42. A. palustris L. Sp. pl. (1753) 846; Karsch. in Kryl. Fl. Zap. Sib. 11 (1949) 2817; Grub. Konsp. fl. MNR [Conspectus of Flora of Mongolian People's Republic] (1955) 267; Poljak. in Fl. SSSR, 26 (1961) 491; Leonova in Grub. Opred. rast. Mong. [Key to Plants of Mongolia] (1982) 246; Fl. Intramong. 6 (1982) 145; Fl. Sin. 76, 2 (1991) 70; Fl. Intramong. ed. 2, 4 (1993) 622; Gub. Konsp. fl. Vneshn. Mong. [Conspectus of Flora of Outer Mongolia] (1996) 98. — Ic.: Fl. Intramong. (1982) tab. 51, fig. 9, 10; Fl. Intramong. (1993) tab. 243, fig. 9, 10.

Described from Siberia. Type in London (Linn.).

In dry valley and coastal solonetzic meadows, rocky and rubbly steppe slopes, pebble beds, solonetzes, gorge floors, glades and borders of forests, along banks of rivers and lakes, sometimes as weed in plantations and fallow lands.

IA. Mongolia: *Khobd.* (Kharkhira mountain group, Burtu area, July 13, 1903 — Gr.-Grzh.), *Mong. Alt.* (Khasargu-Khairkhan, Dundu-Tserengol river valley, Sept. 15, 1930 — Pob.), *Cent. Khalkha* (Dzhirgalante river basin, around spring, Aug. 10, 1925 — Krasch. et Zam.; vicinity of Ikhe-Tukhum-nor lake, July 1926 — Zam.; 5 km south-east of Choiron, along fringe of solonchak lowland, Aug. 22, 1940; 10-15 km eastward of Delger-Khan, on loamy sand, July 1955 — Yun.), *East. Mong.* ("Shilin-Khoto, Khailar, Baotou" — Fl. Intramong. (1993) l.c.), *Depr. Lakes* (around Kobdo town, July 18, 1906 — Sap.; Khan-Khukhei mountain range, feather grass steppe, July 20; south-east. of Bain-nur lake, Berig-Del' sand, July 25–1945, Yun.; 30 km east of Khara-us-nur lake, July 23; 20 km southwest of Ulangom town, in Kharkhira-gol river valley, Aug. 23, 1979 — Gub.), *Val. Lakes* (on right bank of Ongiin river — opposite Orgiin urton, July 27, 1927 — Klem.; Bain-Khongor, east. portion of Tuiliin-Tal area, steppe, Aug. 27, 1943 — Yun.), *Gobi Alt.* (Dundu-Saikhan mountains, sandy bed, July 13, 1909 — Czet.; Ikhe-Bogdo mountain range, on slope, Aug. 12, 1927 — Simukova; Barun-Saikhan mountains, among rock placers, Sept. 20, 1931 — Ik.-Gal.; Ikhe-Bogdo mountain range, Nariin-Khurimt gorge, river estuary, July 30, 1948 — Grub.; Dzun-Saikhan mountain range, on road to Ëlo-ama creek valley from Dalan-Dzadagad, on ravine floor, July 21, 1970 — Grub., Ulzij., Tsetsegbalzhid; Ikhe-Bogdo mountain range, Ulyastai creek valley, coastal pebble beds, July 2, 1972 — Banz-ragch; Gurban-Saikhan mountain range, Khaltyn-Daba pass, in gorge, Aug. 9, 1970 — Sanczir; 15 km east of Delger-Khid, July 8, 1971 — Isach. et Rachk.), Ordos (Taigukhai area, Aug. 29, 1884 — Pot.).

General distribution: West Sib. (Altay), East, Sib. (Sayans), Far East (south), Nor. Mong. (Hent., Hang., Mong.-Daur.), China (Dunbei).

43. A. persica Boiss. Diagn. pl. or. 1, 6 (1846) 91; Pamp. in Nouv. Giorn. Bot. Ital. n.s. 36 (1930) 385; Poljak. in Fl. SSSR, 26 (1961) 506; Ikonn. Opred. rast. Pamira [Key to Plants of Pamir] (1963) 239; Fl. Kirgiz. 11 (1965) 171; Fl. Kazakhst. 9 (1966) 94; Fl. Xizang. 4 (1985) 753; Fl. Sin. 76, 2 (1991) 35; Opred. rast. Sr. Azii (Key to Plants of Mid. Asia) 10 (1993) 552. — *A. togusbulakensis* B. Fedtsch. in Trav. Mus. Bot. Ac. Sci. Petersb. 1 (1902) 143. — *A. persica* var. *togusbulakensis* (b. Fedtsch.) Filat. in Nov. Syst. Herb. Inst. Bot. Ac. Sci. Kasach. 2 (1964) 66. — Ic.: Fl. SSSR, Plate 22, fig. 2; Fl. Kirgiz, Plate 6, fig. 2.

Described from Iran. Type in Geneva (G).

On rocky, rubbly and melkozem slopes, pebble beds, talus, from foothills to upper mountain belt.

IIIB. Tibet: *Chang Tang* ("Zhitu"' — Fl. Xizang. l.c.), *South.* ("Lhasa" — Fl. Xizang. l.c.).

General distribution: East. Pam., Mid. Asia (West. Tien Shan, Pam.-Alay), Fore Asia (Iran).

44. A. phaeolepis Krasch. in Animad. Syst. Herb. Univ. Tomsk, 1, 2 (1949) 3; Krasch. in Kryl. Fl. Zap. Sib. 11 (1949) 2808; Grub. Konsp. fl. MNR [Conspectus of Flora of Mongolian People's Republic] (1955) 267; Poljak. in Fl. SSSR, 26 (1961) 474; Fl. Kazakhst. 9 (1966) 90; Enum. vasc.

pl. Xizang. (1980) 364; Leonova in Grub. Opred. rast. Mong. [Key to Plants of Mongolia] (1982) 247; Fl. Intramong. 6 (1982) 150; Fl. Xizang. 4 (1985) 757; Fl. Sin. 76, 2 (1991) 56; Fl. Intramong. ed. 2, 4 (1993) 610; Opred. rast. Sr. Azii [Key to Plants of Mid. Asia] 10 (1993) 551; Gub. Konsp. fl. Vneshn. Mong. [Conspectus of Flora of Outer Mongolia] (1996) 99. —Ic.: Fl. Intramong. (1982) tab. 53, fig. 9, 10; Fl. Intramong. (1993) tab. 241, fig. 9, 10.

Described from Siberia (Altay). Type in St.-Petersburg (LE).

In larch forests on rocky, rubbly steppe slopes of mountains, in subalpine and alpine mountain belts, in valleys and on river and lake banks, around springs.

IA. Mongolia: *Khobd.* (south. peak of Kharkhira mountain, July 24, 1879 — Pot.; 21 km north of Ulan-Daba pass, on forest border, July 16, 1977; 22 km west of Sagil somon, Shara-Khadny-gol river, larch forest, Aug. 2, 1978; Shagan-Shivet mountain, Sheep's fescue-forbs steppe, July 4, 1978 — Karam.), *Mong. Alt.* (Taishiri-Ola mountain range, in forest, July 15, 1877 — Pot.; east of Ak-Korum pass, alpine meadow, July 13, 1908 — Sap.; 8 km south-east of Tsepeg-nur somon, larch forest, Aug. 9, 1930 — Pob.; Adzhi-Bogdo mountain range, sheep's fescue mountain steppe, Aug. 7; same site, upper course of Indertiin-gol river, subalpine steppe, Aug. 25, 1947, Yun.; 30 km south of Munkh-Khairkhan settlement, upper course of Bodonchiin-gol river, July 25, 1979 — Gub.; upper course of Nariin-gol river, July 15; Shadzgaityn-nuru mountain range, south. slope toward Khoidt-Dzhirgalantyn-gol river, July 27–1984, Kam. et Dar.), *East. Mong.* ("Shilin-Khoto town, Khukh-Khoto [Dadinshan'], Yakeshi town, Ulantsab" — Fl. Intramong. (1993) l.c.), *Gobi Alt.* (Dundu-Saikhan mountains, midbelt, July 7, 1909 — Czet.; Dundu-Saikhan on meadow, 1931 — Ik.-Gal.; Dundu-Saikhan mountain range, mountain steppe, July 22, 1943 — Yun.; Gurban-Saikhan mountains, Élyn-ama creek valley, July 27, 1970 — Sanczir; Ikhe-Bogdo mountain range, forbs mountain steppe, June 29, 1972 — Banzragch).

General distribution: Jung.-Tarb., West. Sib. (Altay), East. Sib. (Sayans), Nor. Mong. (Fore Hubs., Hent., Hang., Mong.-Daur).

45. **A. roxburgiana** Wall. ex Bess. in Bull. Soc. natur. Moscou, 9 (1836) 57; Hemsl. Fl. Tibet. (1902) 57; Enum. vasc. pl. Xizang. (1880) 364; Fl. Xizang. 4 (1985) 766; Fl. Sin. 76, 2 (1991) 104.

Described from Himalayas (Kashmir). Type in London (K). Map 6.

On rocks, pebble beds in river valleys.

83   IIIA. Qinghai: *Nanshan* (Tetung-gol river, Aug. 20, 1872; same site, Kuku-usu river, July 2, 1880 — Przew.; Yuzhno-Kukunor mountain range, Qaidam-gol river, rocks, Sept. 10, 1894 — Rob.).

IIIB. Tibet: *Weitzan* (Dzhagyn-gol river, July 2, 1900 — Lad.); "Chang Tang (Dansyun)" — Fl. Xizang. (1985) l.c.).

General distribution: China (South-West.).

Note. Differs from related species *A. leucophylla* (Turcz. ex Bess). Pamp. and *A. argyi* Levl. et Vaniot in less longer (up to 30 cm) fertile shoots, much smaller lower and middle cauline leaves, with narrow (0.5–2 (3) mm broad) terminal leaf lobules and outer phyllary with light brownish border. Mesoxerophyte, a species derived from mesophytic group *A. vulgaris.*

46. A. rubripes Nakai in Bot. Mag. Tokyo, 31 (1917) 112; Poljak. in Fl. SSSR, 26 (1961) 450; Grub in Novit. syst. pl. vasc. 9 (1972) 296; Leonova in Grub. Opred. rast. Mong. [Key to Plants of Mongolia] (1982) 249; Fl. Intramong. 6 (1982) 137; Fl. Sin. 76, 2 (1991) 115; Fl. Intramong. ed. 2, 4 (1993) 635; Gub. Konsp. Fl. Vneshn. Mong. [Conspectus of Flora of Outer Mongolia] (1996) 99. — *A. venusta* Pamp. In Nouv. Giorn. Bot. Ital. n.s. 36 (1930) 450; Grub. Konsp. fl. MNR [Conspectus of Flora of Mongolian People's Republic] (1955) 269.

Described from Korea. Type in Tokyo (TI).

In forest borders, meadows, along river valleys.

IA. Mongolia: *East. Mong.* (Khalkhin-gol river, July 24; Buirnur lake, on sandy soil, Aug. 18 – 1977, Dar.; "Khailar town, Shilingol ajmaq, Khukh-Khoto town [Dapinshan']" — Fl. Intramong. (1993) l.c.), *Cent. Khalkha, West. Gobi* — Gub. Konsp. fl. Vneshn. Mong. [Conspectus of Flora of Outer Mongolia] (1996).

General distribution: Far East, Nor. Mong. (Hang., Mong.-Daur., Fore Hing.), China (Dunbei), Korean Peninsula.

Note. Species of group *A. vulgaris* L., typical of Manchurian steppe flora. In laciniation and pubescence of leaf blade, *A. rubripes* Nakai is close to *A. vulgaris* but, in form and arrangement of anthodia (elongated, compact, subspicate panicle), this species resembles *A. integrifolia* L. The find of *A. rubripes* in West. Gobi, in our opinion, is extremely unlikely.

47. A. rupestris L. Sp. pl. (1753) 841; Krasch. in Kryl. fl. Zap. Sib. 11 (1949) 2794; Poljak. in Fl. SSSR, 26 (1961) 508, excl. syn. *A. viridis* Willd. ex Bess.; Fl. Kazakhst. 9 (1966) 98; Grub. in Novit. syst. pl. vasc. 9 (1972) 296; Leonova in Grub. Opred. rast. Mong. [Key to Plants of Mongolia] (1982) 269; Fl. Sin. 76, 2 (1991) 17; Opred. rast. Sr. Azii [Key to Plants of Mid. Asia] 10 (1993) 554; Gub. Konsp. fl. Vneshn. Mong. [Conspectus of Flora of Outer Mongolia] (1996) 99.

Described from Gotland (Sweden). Type in London (Linn.).

On rocky and rubbly steppe slopes, on solonchak and solonetzic meadows, along banks of rivers and lakes, in gorges.

IA. Mongolia: *Mong. Alt.* (Khara-Dzarga mountain range, Khair-Khan-Duru river slope, Aug. 25, 1930 — Pob.; upper course of Khara-gaitu-gol river, left tributary of Bulugun river, Aug. 24; valley of Indertiin-gol river, rocky southern slope, Aug. 25 – 1947, Yun.; 30 km south of Munkh-Khairkhan settlement, on slope, July 25; upper course of Dzhasapy river, Aug. 13 – 1979, Grub.; Uinchiin-gol river basin, valley of Arshintyn-gol river, on rocks, Aug. 14, 1979 — Grub.; Munkh-Khairkhan mountain range, 75 km south-west of Ologoi-nur lake, near Ulan-Daba pass, Aug. 12, 1982 — Gub.), *Depr. Lakes* (34 km south of Sagil somon, July 6, 1978 — Karam.).

IIA. Junggar: *Jung. Ala Tau* (Shuvatyn-Daba pass, northward of Sairam-nur lake, mountain steppe, 1957; Dzhair mountain range, on road to Otu settlement from Toli settlement, mountain steppe, Aug. 9, 1957 — Yun. et Yuan'), *Jung. Gobi* (130 km south-east of Bulgan somon, mountain steppe, Aug. 12, 1977 — Volk.).

General distribution: Jung.-Tarb., Nor. Tien Shan; Europe, Mediterr., Mid. Asia, West. Sib. (Altay), East. Sib. (Sayans), Nor. Mong. (Fore Hubs., Hent., Hang., Mong.-Daur.).

48. A. rutifolia Stechm. ex Spreng. Syst. Veg. 3 (1826) 488; Krasch. in Kryl. Fl. Zap. Sib. 11 (1949) 2789; Grub. Konsp. fl. MNR (Conspectus of Flora of Mongolian People's Republic] (1955) 268; Poljak. in Fl. SSSR, 26 (1961) 505; Ikonn. Opred. rast. Pamira [Key to Plants of Pamir] (1963) 239; Fl. Kirgiz. 11 (1965) 168; Fl. Kazakhst. 9 (1966) 102; Leonova in Grub. Opred. rast. Mong. [Key to Plants of Mongolia] (1982) 250; Fl. Xizang. 4 (1985) 753; Fl. Intramong. 8 (1985) 329; Fl. Sin. 76, 2 (1991) 19; Opred. rast. Sr. Azii (Key to Plants of Mid. Asia] 10 (1993) 556; Gub. Konsp. Fl. Vneshn. Mong. [Conspectus of Flora of Outer Mongolia] (1996) 99. — *A. turczaninowiana* Bess. in Nouv. Mem. Soc. natur. Moscou, 3 (1834) 23. — *A. falconeri* Clarke in Hook. f. Fl. Brit. Ind. 3 (1882) 320. — *A. rutifolia* var. *altaica* (Kryl.) Krasch. in Kryl. Fl. Sibir Occ. 11 (1949) 2789; Fl. Intramong. ed. 2, 4 (1993) 604. — Ic.: Fl. SSSR, 26, fig. 2; Fl. Intramong. (1985) tab. 145, fig. 1–6.

Described from Siberia. Type in St.-Petersburg (LE).

On steppe rocky slopes, rocks, talus, pebble bed and sandy banks of rivers.

IA. Mongolia: *Khobd., Mong. Alt., Cent. Khalkha, East. Mong., Depr. Lakes, East. Gobi, Gobi Alt., West. Gobi, Alash. Gobi* ("West. Alashan" — Fl. Intramong. l.c.).

IB. Kashgar: *Nor.* (Uch-Turfan, June 27, 1908 – Divn.), *West.* (Baikurt settlement, 83 km north-west of Kashgar on road to Torugart, June 19, 1959 – Yun. et Yuan'), *South.* (south. slope of Russky mountain range, on pebble bed in Aksu river valley, July 1, 1895 – Rob.).

IIA. Junggar: *Jung. Alat.* (Aksu, June 31, 1977 – Fet.; southwest. extremity of Maili mountain range, 20 km north-east of Junggar gateway, rocky slopes, Aug. 14, 1957 – Yun., Li et Yuan'), *Tien Shan* (Dzhagastai, July 23; Sairam lake, Aug. – 1877, A. Reg.; Khaidyk-gol river, tributary of Khor-Teryn, on rocks, Aug. 10, 1893 – Rob.; nor. slopes toward Manas river, valley of Ulan-usu river, at its confluence with Dzhartas, rocky slopes, Aug. 18, 1957 – Yun., Li, Yuan'), *Jung. Gobi* (Baitag-Bogdo mountain range, nor. slopes, Sept. 18, 1848 – Grub.; same site, on rocks, Aug. 2, 1979 – Gub.; Bulgan river basin, 1 km from Ded-Nariin-gol estuary, on rocks, Aug. 17, 1979 – Grub., Muld., Dar.).

IIIB. Tibet: *Chang Tang* (Przewalsky mountain range, on rocks, Aug. 20, 1895 – Rob.).

IIIC. Pamir (Kara-Chukur and Ilik-su, July 16, 1901 – Alekseenko; Pas-Rabat, rocky placer, July 3, 1909 – Divn.).

General distribution: Jung.-Tarb., Nor. and Cent. Tien Shan, Pam.-Alay, Fore Asia, Mid. Asia, West. and East. Sib., Far East, Nor. Mong. (Hent., Hang., Mong.-Daur.), China (Nor., Nor.-West.), Himalayas (West.).

49. A. santolinifolia (Turcz. ex Pamp.) Krasch. in Kryl. Fl. Zap. Sib. 11 (1949) 2791; Grub. Konsp. fl. MNR [Conspectus of Flora of Mongolian People's Republic] (1955) 268; Poljak. in Fl. SSSR, 26 (1961) 465; Ikonn. Opred. rast. Pamira [Key to Plants of Pamir] (1963) 237; Fl. Kirgiz. 11 (1965) 163; Fl. Kazakhst. 9 (1966) 91; Enum. vasc. pl. Xizang. (1980) 364; Leonova in Grub. Opred. rast. Mong. [Key to Plants of Mongolia] (1982) 247; Fl. Intramong. 8 (1985) 322; Fl. Xizang. 4 (1985) 756; Opred. rast. Sr. Azii [Key to Plants of Mid. Asia] 10 (1993) 551; Gub. Konsp. fl. Vneshn.

Mong. [Conspectus of Flora of Outer Mongolia] (1996) 99.—*A. santolinifolia* var. *stepposa* Darijmaa in Bull. Soc. natur. Mosc. 97, 5 (1992) 67.—*A. sacrorum* Ledeb. var. *minor* Ledeb. Fl. Alt. 4 (1833) 72.—*A. gmelinii* Web. ex Stechm. var. *turczaninoviana* Bess. in Nouv. Mem. Soc. natur. Moscou, 3 (1834) 87.—*A. sacrorum* var. *santolinifolia* Pamp. in Nouv. Giorn. Bot. Ital. n.s. 34 (1927) 603; Fl. Intramong. ed. 2, 4 (1993) 613.—*A. sacrorum* auct. non Ledeb.: Franch. Pl. David. 1 (1884) 170, p.p.; Forbes et Hemsl. Index Fl. Sin. 1 (1888) 444; Hemsl. Fl. Tibet. (1902) 183, p.p.; Pamp. Fl. Carac. (1930) 211, p.p.; Kitag. Lin. Fl. Mansh. (1939) 430. —Ic.: Ledeb. Ic. pl. fl. ross. 5 (1834) tab. 471; Fl. Kazakhst. Plate 10, fig. 3; Grub. Opred. rast. Mong. [Key to Plants of Mongolia] Plate 83, fig. 608; Fl. Intramong. (1985) tab. 141, fig. 6.

Described from East. Siberia. type in St.-Petersburg (LE).

On rocky slopes of mountains, talus, in desert-steppe valleys of rivers on pebble beds, sandy banks of rivers and lakes, ascending up to forest belt.

IA. Mongolia: *Khobd.* (Kharkira river valley, July 21, 1879—Pot.; Boro-Borgosun river, tributary of Saksai river, July 2, 1903—Sap.; 10 km from Tsagan-nur lake, on road to Bayan-Ulegei, Aug. 3; 10–12 km south of Bayan-Ulegei on road to Tolbo-nur, Aug. 4 1945, Yun.), *Mong. Alt.* (near Khobdo, rocky steppe, Aug. 1, 1899—Lad.; Taishiri-ula mountain range, among rocks, Aug. 15; Khalyun river terrace, Aug. 16; Khara-Adzarga mountain range, valley of Sakhir-Sala river, pebble bed, Aug. 21; Bodonchiin-gol river at exit from gorge, Sept. 23–1930, Pob.; Bayan-Undur somon, on mountain, Aug. 20, 1943; 25 km west of Tsagan-01 lake on road to Kobdo, Aug. 28, 1944—Yun.; Taishir-ula, on floor of sandy gorge, July 11, 1945; 2–3 km south of Tamcha lake, July 17, 1947; Adzhi-Bogdo mountain range, Aug. 1947—Yun.; 20 km south-east of Bayan-Undur somon centre, on slopes and floor of gorge, July 20, 1948—Grub.; Khasagtu-Khairkhan mountain range, 30 km south of Dzhargalant somon centre, Aug. 14, 1947—Yun.; nor. bank of Dayan-nur lake, Aug. 8–9, 1972—Metel'tseva; 22 km south-west of Bugat somon centre, Sept. 2, 1973—Isach. et Rachk.; Adzhi-Bogdo mountain range, talus, Aug. 1977— Grub.; 30 km south of Munkhu-Khairkhan settlement. on slopes, July 25, 1979—Gub.), *Cent. Khalkha* (Dzhargalante river basin, terrace, Sept. 13, 1925—Krasch. et Zam.), *Depr. Lakes* (Bogden-gol river, Aug. 28, 1879—Pop.; south. bank of Khara-nur, Aug. 18, 1944— Yun.; 30 km south-east of south. bank of Khara-us lake, July 24, 1979—Gub.; 40 km east-south-east of Undurkhangai, Aug. 16, 1979—Karam.), *Val. Lakes* (rocky slopes toward Tatsin-gol river, Sept. 17, 1924—Pavl.), *Gobi Alt.* (Arts-Bogdo mountain range, 1925; Gurban-Saikhan mountain range, 1925—Chaney; Dundu-Saikhan mountain range, rocky slopes, Aug. 17, 1931—Ik.-Gal.; Gurban-Saikhan mountain range, rocky slopes, July 22; Ikhe-Bogdo mountain range, Sept. 7–1943 Yun.; same site, southeast. slope of Narin-Khurimt gorge, on rocks, July 28, 1948—Grub.; Nemegetu-ula mountain range, rocky slope, July 29, 1972—Guricheva; Barun-Saikhan mountain range, Aug. 23, 1953— Dashnyam; Ikhe-Bogdo mountain range, July 14, 1979—Isach.), *East. Mong.* ("Shilin-Khoto, Ulantsab, Khukh-Khoto, Baotou, Ulan'shan' "—Fl. Intramong. (1993) l.c.), *Khesi* (mountains near Tszyutsyuan' town, Sept. 2, 1958—Lavr.).

IC. Qaidam: *Plain* (on Nomokhun-gol river, nor. foothill of Burkhan-Budda mountains, Aug. 14, 1884—Przew.), *Mount.* (mountains, in juniper groves, Aug. 3, 1901—Lad.).

IIA. Junggar: *Cis-Alt.* (20 km nor. of Kok-Togoi, right bank of Kairta river, valley of Kuidun river, Aug. 15, 1959 – Yun.), *Tien Shan* (Khaidyk-gol river, Aug. 6, 1893 – Rob.; vicinity of Urumchi town, Sept. 25, 1929 – Pop.; Boro-Khoro mountain range, on highway to Kul'dzha from Urumchi, Aug. 19; Ketmen' mountain range, 3-4 km above Sarbushin settlement on road to Kyzyl-Kure from Ili, talus, Aug. 23 – 1957, Yun.; vicinity of Urumchi, spruce forest, Aug. 29, 1959 – Petr.), *Jung. Gobi* (Barlagiin gol river, on road to Bodonchiin-Baishing, pebble bed, Sept. 9; Baktak-Bogdo mountain range, slopes to Ulyasutai-gol gorge, Sept. 18 – 1948, Grub.).

IIIA. Qinghai: *Nanshan* (15 km south of Aksai settlement, Altyn-tag mountain range, rocky slopes, Aug. 2, 1958 – Petr.).

IIIB. Tibet: *Chang Tang* ("Zhitu, Gaitsze" – Fl. Xizang. l.c.), *South.* ("Nan'mulin, Pulan', Chzhada" – Fl. Xizang. l.c.).

IIIC. Pamir (Kenkol gorge, near Togai-Bashi village, July 30, 1913 – Knorr.; Kunlun' mountain range, Pasrabat area, Aug.-Sept.; Tekenlik pass, Aug. – 1944, Serp.; Kunlun' mountain range, lateral arid right bank of valley toward Tiznaf river, on slopes, July 4; same site; King-tau mountain range, rocky steppe slopes, July 10 – 1959, Yun.).

General distribution: Fore Balkh., Jung.-Tarb., Nor. and Cent. Tien Shan, East. Pam., West. and East., Sib., Far East, Nor. Mong. (Hang., Mong.-Daur.), China (Nor., Nor.-West. Cent.).

50. **A. selegensis** Turcz. ex Bess. in Nouv. Mem. Soc. natur. Moscou, 3 (1834) 50; Poljak. in Fl. SSSR, 26 (1961) 454; Leonova in Grub. Opred. rast. Mong. [Key to Plants of Mongolia] (1982) 248; Fl. Intramong. 6 (1982) 135; Fl. Sin. 76, 2 (1991) 144; Fl. Intramong. ed. 2, 4 (1993) 634; Gub. Konsp. fl. Vneshn. Mong. [Conspectus of Flora of Outer Mongolia] (1996) 99. – *A. selengensis* var. *shansiensis* Y.R. Ling in Bull. Bot. Res. 8, 3 (1988) 5; Fl. Intramong. ed. 2, 4 (1993) 634. – *A. vulgaris* L. var. *selengensis* (Turcz. ex Bess.) Maxim. in Bull. Ac. Sci. Petersb. 8 (1872) 536. – Ic.: Fl. Intramong. 6 (1982) tab. 49, fig. 8, 9.

Described from Siberia (Selenga river). Type in St.-Petersburg (LE).

In forest glades and borders, floodplain meadows, along river banks, in willow thickets along river valleys.

IA. Mongolia: *East. Mong.* ("Shilingol' ajmaq, Khukh-Khoto town" – Fl. Intramong. 1993, l.c.).

General distribution: East. Sib. (Sayans), Far East, Nor. Mong. (Hang., Mong.-Daur., Fore Hing.), China (Dunbei), Korean peninsula.

Note. Forest species of group *A. vulgaris* L. with stable morphological characteristics.

51. **A. sericea** Web. ex Stechm. Artemis. (1775) 16; Krasch. in Kryl. Fl. Zap. Sib. 11 (1949) 2795; Grub. Konsp. Fl. MNR [Conspectus of Flora of Mongolian People's Republic] (1955) 268; Poljak. in Fl. SSSR, 26 (1961) 501; Fl. Kazakhst. 9 (1966) 101; Leonova in Grub. Opred. rast. Mong. [Key to Plants of Mongolia] (1982) 249; Fl. Intramong. 6 (1982) 158; Fl. Sin. 76, 2 (1991) 11; Fl. Intramong. ed. 2, 4 (1993) 603; Opred. rast. Sr. Azii [Key to Plants of Mid. Asia] 10 (1993) 556; Gub. Konsp. fl. Vneshn. Mong. [Conspectus of Flora of Outer Mongolia] (1996) 96. – Ic.: Gmel. Fl. Sib.. 2 (1749) 131, tab. 64, fig. 1; Fl. Intramong. (1993) tab. 55, fig. 7, 8.

Described from Siberia. Gmelin's cited sketch serves as type.

On steppe rubbly and rocky slopes of low mountains, borders of birch forests, rarely on weakly saline meadows.

IA. Mongolia: *Cent. Khalkha* (basin of Dzhargalante river basin, head of Kharukhe river, Uste mountains, Sept. 10, 1925 – Krasch. et Zam.), *East. Mong.* (Boro-Khurkhe, Aug. 25, 1927 – Terekhovko; "Yakeshi, Manchuria town" – Fl. Intramong. (1993) l.c.).

General distribution: Fore Balkh. (nor.-east. section), Europe, West. and East. Sib., Nor. Mong. (Hent., Hang., Mong.–Daur.).

52. **A. sieversiana** Willd. Sp. pl. 3, 3 (1803) 1845; Franch. Pl. David. 1 (1884) 171; Forbes et Hemsl. Index Fl. Sin. 1 (1888) 445; Pamp. Fl. Carac. (1930) 213; Kitag. Lin. Fl. Mansh. (1939) 433; Walker in Contribs. U.S. Nat. Herb. 2, 8 (1941) 667; Krasch. in Kryl. Fl. Zap. Sib. 11 (1949) 2818; Grub. Konsp. fl. MNR [Conspectus of Flora of Mongolian People's Republic] (1955) 269; Poljak. in Fl. SSSR, 26 (1961) 103; Fl. Kirgiz. 11 (1965) 172; Fl. Kazakhst. 9 (1966) 103; Enum. vasc. pl. Xizang. (1980) 364; Leonova in Grub. Opred. rast. Mong. [Key to Plants of Mongolia] (1982) 246; Fl. Intramong. 6 (1982) 155; Fl. Xizang. 4 (1985) 1744; Pl. Helansch. (1986) 357; Fl. Sin. 76, 2 (1991) 9; Fl. Intramong. ed. 2, 4 (1993) 600; Opred. rast. Sr. Azii [Key to Plants of Mid. Asia] 19 (1993) 557; Gub. Konsp. fl. Vneshn. Mong. [Conspectus of Flora of Outer Mongolia] (1996) 99. — Ic.: Fl. Kazakhst. 9, Plate 11, fig. 3; Grub. Opred. rast. Mong. [Key to Plants of Mongolia] Plate 134, fig. 613; Fl. Intramong. (1982) tab. 53, fig. 11–13; Fl. Xizang. (1985) tab. 321, fig. 1–6; Fl. Intramong. (1993) tab. 236, fig. 1–6.

Described from Siberia. Type in Berlin (B).

On steppe rocky and rubbly slopes of mountains, solonetzic meadows, on banks of rivers and lakes, floors of gorges, often as weed around residences, roadsides, farms, fallow lands, near wells and springs.

IA. Mongolia: *Khobd., Mong. Alt., Cent. Khalkha, East. Mong., Depr. Lakes, Val. Lakes, Gobi Alt., East. Gobi, West. Gobi, Alash. Gobi, Ordos.*

IB. Kashgar: *Nor.* (Uch-Turfan, July 27, 1908 – Divn.; Yakka-aryk village on Muzart-Dar'ya, Aug. 13; Kyzyl-su river basin, beyond Kashgar, Kara-Tash river floodplain, Aug. 20–1929, Pop.), *South.* (on Nura river up to Gendzhi-Dar'ya, July 1; on Keriya river, along Nura river, July 23–1885, Przew.; nor. foothills of Kuen'-Lun' foothills, Sept. 13, 1889 – Rob.).

IC. Qaidam: *Plain* (Nomozhun-gol river, downstream, Aug. 4; Khatu-gol river, in lower belt of Burkhan-Budda mountain range, Aug. 12–1884, Przew.).

IIA. Junggar: *Jung. Alat.* (left bank of lower Borotala, Aug. 22, 1878 – A. Reg.; Dzhair mountain range, intermontane plane on road to Chipeitszy-Chuguchak, Aug. 5, 1951 – Mois.; same site, 1–2 km north of Otu settlement, on road to Chuguchak from Aktash, in chee grass thickets, Aug. 4, 1957 – Yun., Li, Yuan'), *Tien Shan* (Ili river basin, Aug. 1876 – Przew.; Arystan on Ili river, Aug. 7, 1878 – Fet.; Urtak-Saras, Aug. 14, 1878 – A. Reg.; Bain-gol river, right tributary of Tekes river, on meadow, June 29, 1893 – Rob.; 3

km east of Nyutsuan'tszy settlement on road to Ulan-usu river, fallow land, July 24; Ketmen' mountain range, 8–10 km south of Sarbushin settlement on road to Kzyl-Kure from Kul'dzha, on slopes, Aug. 23–1957, Yun., Li et Yuan'), *Jung. Gobi* (lower course of Uinchi-gol river 15–20 km before winter camp of somon, solonchak meadow, July 29, 1947 – Yun.; Baitag-Bogdo mountain range, Ulyastugol gorge, in floodplain, Sept. 18, 1948 – Grub.).

IIIA. Qinghai: *Nanshan* (road along Tetung river valley, Aug. 28, 1872; same site, July 15, 1880 – Przew.; Van-tszi-tszin' village, on irrigation ditch, Aug. 16, 1908 – Czet.; 60 km south-east of Chzhan'e town, high foothills of Nanshan, July 12; Tetung river valley, slopes of conical hillocks near stud farm, Aug. 20–1958; Xining, 5 km south of Sadtszana, meadow in river floodplain, Aug. 6, 1959 – Petr.).

IIIB. Tibet: *South.* ("Lhasa, Nan'mulin" – Fl. Xizang. (1985) l.c.).

IIIC. Pamir (Pas-Rabat river, slopes of mountains in Toili-Bulun area, Aug. 2, 1909 – Divn.; Issyk-su river estuary, July 19; Kulan-Aryk area, between Zad settlement and Tashui river, Aug. 20; Pakhtu river, Aug. 22–1942, Serp.; 2 km west of Tash-Kurgan, fallow land, June 13, 1959 – Yun., Li et Yuan').

General distribution: Aralo-Casp., Fore Balkh., Jung.-Tarb., Nor. and Cent. Tien Shan, East. Pamir, Europe, West. and East. Sib., Far East, Nor. Mong., China (Nor., Nor.-West., Cent.), Himalayas (Kashmir).

53. **A. stolonifera** (Maxim.) Kom. in Fl. Mansh. 3 (1907) 636; Poljak. in Fl. SSSR, 26 (1961) 446; Fl. Sin. 76, 1 (1991) 85; Fl. Intramong. ed. 2, 4 (1993) 623. – *A. vulgaris* L. var. *stolonifera* Maxim. Prim. Fl. Amur. (1851) 161. – *A. integrifolia* L. var. *stolonifera* (Maxim.) Pamp. in Nouv. Giorn. Bot. Ital. n.s. 36 (1930) 481.

Described from Far East (Ussuri). Type in St.-Petersburg (LE).

In forest zone in meadows, on slopes of conical hills, river valleys.

IA. Mongolia: *East. Mong.* ("Khulunbuir ajmaq, Yakeshi" – Fl. Intramong. (1993) l.c.).

General distribution: Far East (Zee-Bur.: south, Udsk., Ussur., Sakh.), Japan, China (Dunbei).

Note. A typical species with entire leaf blade sharp-toothed on margin; so far, this is lacking in our collection from Mongolian territory and cited after Fl. Intramong. l.c.

54. **A. stracheyi** Hook. f. et Thoms. ex Clarke, Comp. Ind. (1876) 164; Hemsl. vasc. pl. Xizang. (1980) 364; Fl. Xizang. 4 (1985) 759; Fl. Sin. 76, 2 (1991) 29. – Ic.: Fl. Xizang. (1985) tab. 328.

Described from Himalayas (Kashmir). Type in London (K).

On slopes of mountains, in rock crevices, 3400–4000 m alt.

IC. Qaidam: *Plain* (Ikhe-Qaidamin-nor lake, July 13, 1895 – Rob.).

IIIA. Qinghai: *Nanshan* (24 km from Aksai settlement on road to Qaidam, pass through Altyn-Tag mountain range, Aug. 2, 1958 – Petr.).

IIIB. Tibet: *Chang Tang* (Zhitu, Getszy, Shuankhu – Fl. Xizang. (1985) l.c.), *South* (Tszyantszy, Sage, Pulan', Chzhunba" – Fl. Xizang. (1985) l.c.).

General distribution: Himalayas (Kashmir).

Note. Species close to *A. campbellii* Hook. f. et Thoms. ex Clarke.

55. A. subulata Nakai in Bot. Mag. Tokyo, 29 (1915) 18; Poljak. in Fl. SSSR, 26 (1961) 450; Fl. Intramong. 6 (1982) 129; Fl. Sin. 76, 2 (1991) 126; Fl. Intramong. ed. 2, 4 (1993) 629; Gub. Konsp. fl. Vneshn. Mong. [Conspectus of Flora of Outer Mongolia] (1996) 99. — *A. integrifolia* L. var. *subulata* (Nakai) Pamp. in Nouv. Giorn. Bot. Ital. n.s. 36 (1930) 480. — Ic.: Fl. Intramong. (1993) tab. 245, fig. 6, 7.

Described from Korea. Type in Tokyo (TI).

In forest zone in meadows, among shrubs.

IA. Mongolia: *East. Mong.* (Khuntu somon, 65 km west of Togë-gol river, forbs meadow, Aug. 7, 1949 — Yun.; Sartu station, meadow, Sept. 12, 1951 — Chaj Ta-Chang; "Shilin-khoto, Ulantsab, Khukh-khoto town [Datsin'shan']" — Fl. Intramong. (1993) l.c.).

General distribution: Far East, Nor. Mong. (Fore Hubs., Fore Hang.), China (nor.), Korean peninsula.

56. A. succulentoides Ling et Y.R. Ling in Acta Phytotax. Sin. 18 (1980) 504; Enum. vasc. pl. Xizang. (1980) 364; Fl. Xizang. 4 (1985); Fl. Sin. 76, 2 (1991) 34. — Ic.: Fl. Xizang. (1985) tab. 322, fig. 1–11.

Described from Tibet. Type in Beijing (PE).

In cereal grass steppes at 3800 m alt.

IIIB. Tibet: *South.* (Lhasa town region, Aug. 16, 1965 — Chzhan Yun-Tyan' et Lai Kai-Yun, typus!).

General distribution: endemic.

Note. Species related to *A. anethoides* Mattf.

57. A. superba Pamp. in Nouv. Giorn. Bot. Ital. n.s. 34 (1930) 473; Leonova in Grub. Opred. rast. Mong. [Key to Plants of Mongolia] (1982) 248; Gub. Konsp. fl. Vneshn. Mong. [Conspectus of Flora of Outer Mongolia] (1996) 99.

Described from China. Type in Firenze (Florence) (FI).

On rocky slopes of low mountains, sandy banks and pebble beds of rivers and lakes.

IA. Mongolia: *East. Mong.* (Tsen-khariin-gol river, 28 km north-east of Delgerkhan somon, pebble bed in river floodplain, Aug. 8; 12 km east of Shilin-Bogdo-ula somon, on rocky slopes, Aug. 15; 3 km north of Dzhargalant post, on rocky slopes, Aug. 16–1980, Dar.).

General distribution: Nor. Mong. (Hang., Hent.), China (Dunbei).

Note. Exhibits extremely close genetic affinity with *A. verlotiorum* Lamotte and *A. vulgaris* L. but this species is most xerophilized with several stable morphological characteristics: least sizes of vegetative and regenerative organs, as well as more compact pubescence of the plant as a whole.

58. A. sylvatica Maxim. Mem. Pres. Acad. Petersb. Div. Sav. 9 (1859) 161 [Prim. Fl. Amur]; Kitag. Lin. Fl. Mansch. (1939) 434; Poljak. in Fl. SSSR, 26 (1961) 445; Leonova in Grub. Opred. rast. Mong. [Key to Plants of Mongolia] (1982) 249; Fl. Intramong. 6 (1982) 132; Fl. Sin. 76, 2 (1991)

149; Fl. Intramong. ed. 2, 4 (1993) 637; Gub. Konsp. Fl. Vneshn. Mong. [Conspectus of Flora of Outer Mongolia] (1996) 99. —Ic.: Fl. Intramong. (1982) tab. 47, fig. 7, 8; Fl. Intramong. (1993) tab. 248, fig. 7, 8.

89 Described from Far East (Ussuri river estuary). Type in St.-Petersburg (LE).

In deciduous, coniferous-deciduous forests, along river valleys, banks of rivers.

IA. Mongolia: *East. Mong., Depr. Lakes, West. Gobi* (Gub. l.c.).

General distribution: Far East, Nor. Mong. (Fore Hing., Mong.-Daur.), China (Nor.), Korean peninsula.

Note. Distinct species with stable morphological characteristics; easily distinguished from all species close to *A. vulgaris* L. in fine anthodia loosely arranged on long, very slender highly declinate branchlets of panicles.

59. **A. tanacetifolia** L. Sp. pl. (1753) 848; Krasch. in Kryl. Pl. Zap. Sib. 11 (1949) 2806; Grub. Konsp. Fl. MNR [Conspectus of Flora of Mongolian People's Republic] (1955) 266; Poljak. in Fl. SSSR, 26 (1961) 468; Leonova in Grub. Opred. rast. Mong. [Key to Plants of Mongolia] (1982) 248; Fl. Sin. 76, 2 (1991) 57; Fl. Intramong. ed. 2, 4 (1993) 614; Gub. Konsp. fl. Vneshn. Mong. [Conspectus of Flora of Outer Mongolia] (1996) 99.—*A. macrobotrys* Ledeb. Fl. Alt. 4 (1833) 73. —Ic.: Fl. Intramong. (1993) tab. 241, fig. 1–6.

Described from Siberia. Type in London (Linn.).

In light coniferous and deciduous forests, in forest borders and glades, in willow thickets, rocky slopes in forests.

IA. Mongolia: *Mong. Alt., Cent. Khalkha, Depr. Lakes* (Gub. l.c.), *Khobd.* (Kharkhira river, July 9, 1879—Pot.), *East. Mong.* ("Khailar town, Shilin-Khoto, Ulantsab, Khukh-Khoto, Ulashan' " —Fl. Intramong. (1993) l.c.), *Alash. Gobi* ("Alashan' (Khelanshan')"—Fl. Intramong. (1993) l.c.).

General distribution: Volga-Kam., East. Sib., Far East, Nor. Mong.. (Fore Hubs., Hent., Hang., Mong.-Daur., Fore Hing.), China.

60. **A. tangutica** Pamp. in Nouv. Giorn. Bot. Ital. n.s. 36 (1930) 426; Enum. vasc. pl. Xizang. (1980) 365; Fl. Xizang. 4 (1985) 773; Fl. Sin. 76, 2 (1991) 155.—*A. vulgaris* L. var. *tangutica* Maxim. in Herb. —Ic.: Fl. Xizang. (1985) tab. 335, fig. 1–4.

Described from Nor.-West. China (Gansu). Lectotype in St.-Petersburg (LE). Plate V. fig. 3.

On clayey-rocky slopes, rocks, gorge floors.

IIIA. Qinghai: *Nanshan* (on Tetung-gol river, 2750 m above sea, Aug. 24, 1880—Przew., lectotypus!; Cherten-ton temple, on sand, Sept. 8, 1901—Lad., syntypus!).

General distribution: endemic in Central Asia.

Note. Differs from proximate species in coriaceous leaf blade glabrous and non-pubescent on both sides. Differs from *A. umbrosa* (Bess.) Pamp. in broad, usually even more deeply lobately incised leaf blades and, from *A. leucophylla* (Turcz. ex Bess.) Pamp., in campanulate anthodia gathered in compact spicate inflorescences on lateral branchlets of narrow-pyramidal panicle.

61. A. umbrosa (Bess.) Pamp. in Nouv. Giorn. Bot. Ital. n.s. 36 (1930) 448; Grub. Konsp. fl. MNR [Conspectus of Flora of Mongolian People's Republic] (1955) 269; Poljak. in Fl. SSSR, 26 (1961) 453; Leonova in Grub. Opred. rast. Mong. [Key to Plants of Mongolia] (1982) 249; Fl. Intramong. 6 (1982) 135; Gub. Konsp. fl. Vneshn. Mong. [Conspectus of Flora of Outer Mongolia] (1996) 99. — *A. vulgaris* L. var. *umbrosa* Bess. in Nouv. Mem. Soc. natur. Moscou, 3 (1834) 52. — *A. lavandulifolia* auct. non DC.: Fl. Intramong. ed. 2, 4 (1993) 625. — Ic.: Fl. Intramong. (1982) tab. 49, fig. 10, 11.

Described from Siberia (Salenga river). Type in St.-Petersburg (LE). Map 5.

In forest grasslands, coastal wet and steppified meadows, among shrubs.

IA. Mongolia: *East. Mong.* (Khalkhin-gol river, meadow, Aug. 7, 1899 — Pot.; Sobinur area, steppe, Sept. 4, 1928 — Tug.; 25 km west of Dariganga town, near spring, Sept. 9, 1931 — Pob.; 38 km south-east of Choibalsan, forbs meadow steppe, Aug. 6; Khalkhingol river, 13 km south-east of Khamar-Daba village, Aug. 11–1949, Yun.).

General distribution: East. Sib. (Sayans), Far East, Nor. Mong. (Mong.-Daur.), China (Nor.), Korean peninsula.

Note. Distinct species with stable characteristics, related to *A. landulifolia* DC. extensively distributed in North-East. China; close to *A. mongolica* (Bess.) Fisch. et Mey. ex Nakai in blade structure.

62. A. tournefortiana Reichb. Iconogr. Bot. Exot. 1 (1824) 6; Poljak. in Fl. SSSR, 26 (1961) 489; Fl. Kirgiz. 11 (1965) 163; Fl. Kazakhst. 9 (1966) 95; Leonova in Grub. Opred. rast. Mong. (Key to Plants of Mongolia] (1982) 247; Fl. Xizang. 4 (1985) 724; Fl. Sin. 76, 2 (1991) 67; Opred. rast. Sr. Azii (Key to Plants of Mid. Asia] 10 (1993) 553; Gub. Konsp. fl. Vneshn. Mong. [Conspectus of Flora of Outer Mongolia] (1996) 99. — Ic.: Reichb. l.c. tab. 5; Fl. Xizang. tab. 326.

Described from the "East". Type in Berlin (B).

Weed along river banks, floors of solonchak valleys, in plantations, pastures, roadsides and around residences.

IA. Mongolia: *East. Mong.* (35 km north-east of Enger-Shanda, on solonetzic soil, Aug. 17, 1949 — Yun.), *East. Gobi* (30 km west of Sain-Shanda town, Barun-Tugrik area, Sept. 17, 1940 — Yun.).

IIA. Junggar: *Jung. Gobi* (lower course of Borotala river, left bank, Aug. 22, 1878 — A. Reg.; Bulugun river floodplain, at wintering site of somon, July 28; lower course of Uinchi-gol river, 15–20 km before wintering site of somon, solonetzic meadow around streams, July 29 — 1947, Yun.).

IIIB. Tibet: *Chang Tang* ("Gaitsze" — Fl. Xizang. l.c.), *South.* ("Nan'mulin" — Fl. Xizang. l.c.).

General distribution: Fore Balkh., Jung.-Tarb., Nor. and Cent. Tien Shan, Fore Asia, Caucasus, Mid. Asia.

63. A. verbenacea (Kom.) Kitag. In Lin. Fl. Mansh. (1939) 434; Fl. Intramong. 8 (1985) 323; Fl. Sin. 76, 2 (1991) 113; Fl. Intramong. ed. 2, 4 (1993) 633. — *A. vulgaris* L. var. *verbenacea* Kom. Fl. Mansh. 3 (1907) 673. — *A. mongolica* (Bess.) Fisch. et Mey. ex Nakai var. *verbenacea* (Kom.) Pamp. in Nouv. Giorn. Bot. Ital., n.s. 36, 4 (1930) 412; Fl. Intramong. 6 (1982) 140. — Ic.: Fl. Intramong. (1985) tab. 142, fig. 1–5; Fl. Intramong. (1993) tab. 247, fig. 1–3.

Described from Manchuria. Type in St.-Petersburg (LE)..

On forest borders, rivers banks.

IA. Mongolia: *East. Mong.* ("Datsinshan" — Fl. Intramong. (1985) l.c.; "Ulantsab, Shilin-gol" — Fl. Intramong. (1993) l.c.), *Alash. Gobi* ("East. Alashan, Helanshan' " — Fl. Intramong. (1993) l.c.).

General distribution: China (Dunbei), Korean peninsula.

Note. In the works cited above, this species was regarded as a synonym of *A. vulgaris* L. or of *A. mongolica* (Bess.) Fisch. et Mey. ex Nakai. A study of the type specimen showed that *A. verbenacea* (Kom.) Kitag. is affiliated to *A. rubripes* Nakai and is probably also a representative of Manchurian flora. Not having herbarium specimens from the regions studied, but considering the extremely large distribution range reported in Fl. Intramong. (1985, 1993), we conformed to the cited references in our treatment of this species.

64. A. verlotiorum Lamotte in Mem. Asso-Fran. Cong. Clerm. (1876) 511; Pamp. in Nouv. Giorn. Bot. Ital. n.s. 36 (1930) 465; Fl. Sin. 76, 2 (1991) 94; Fl. Intramong. ed. 2, 4 (1993) 639.

Described from China. Type in Firenze (Florence) (FI).

On slopes of conical hillocks, meadows, river valleys.

IA. Mongolia: *East. Mong.* ("Shilin-Khoto" — Fl. Intramong. l.c.).

General distribution: China (Dunbei), Japan (nor.), Korean peninsula.

Note. This highly characteristics species from group *A. vulgaris* L. represents Manchurian flora. In laciniation and pubescence of lower and middle cauline leaves, this species is close to *A. umbrosa* (Bess.) Pamp. but, in the form of anthodia and their arrangement on panicle branchlets, to the Kamchatka species *A. opulenta* Pamp.

65. A. vestita Wall. ex DC. Prodr. 6 (1838) 106; Wall. Cat. (1828) No. 3301, nom. nud.; Forbes et Hemsl. Index Fl. Sin. 1 (1888) 445; Kitag. Lin. Fl. Mansh. (1939); Enum. vasc. pl. Xizang. (1980) 319; Fl. Xizang. 4 (1985) 757; Fl. Intramong. 8 (1985) 170; [Drevesn. rast. Tsinkhaya— Wooded Plants of Qinghai] (1987) 634; Fl. Sin. 76, 2 (1991) 49. — *A. sacrorum* auct. non Ledeb.: Franch. Pl. David. 1 (1884) 170; Hemsl. etc.

Described from Himalayas (Kashmir). Type in Geneva (G). Isotype in St.-Petersburg (LE).

On rocky, rubbly and stony slopes of mountains.

IA. Mongolia: *Gobi Alt.* (Barun-Saikhan mountains, rocky slopes, Aug. 23, 1970 — Sanczir).

IIIA. Qinghai: *Nanshan* (Yushno-Kukunorsky mountain range, Ara-gol river, Ikhe-Bilgir area, June 27; same site, Qaidam-gol river, rocks, Sept. 9–1894, Rob.; 66 km west of Xining town, rocky slopes, Aug. 5, 1959 — Petr.).

IIIB. Tibet: *South* ("Lhasa, Nan'mulin, Shigetse" — Fl. Xizang (1985) l.c.).

General distribution: China (Nor.-West., Cent.), Himalayas (West. Kashmir).

66. A. vexans Pamp. in Nouv. Giorn. Bot. Ital. n.s. 36 (1930) 437; Enum. vasc. pl. Xizang. (1980) 365; Fl. Xizang. 4 (1985) 777; Fl. Sin. 76, 2 (1991) 155.

Described from Tibet. Type in St.-Petersburg (LE).

In alpine steppes, valleys of mountain rivers, sometimes on talus.

IIIB. Tibet: *Weitzan* (Dzhagyn-gol river, Mekong river basin, Bar-chyu river, 3600–3900 m above sea, Oct. 1900 — Lad. typus!).

General distribution: endemic.

Note. This alpine mesoxerophyte close to *A. obscura* Pamp. differs in short-petiolate lower and middle cauline leaves and much larger anthodia, usually singly or in loose groups of 2–3 on lateral branchlets of narrow panicle.

67. A. viridis Willd. Sp. pl. 3, 3 (1803) 1829; Krasch. in Kryl. Fl. Zap. Sib. 11 (1949) 2794; Grub. Konsp. fl. MNR [Conspectus of Flora of Mongolian People's Republic] (1955) 251; Ikonn. Opred. rast. Pamira [Key to Plants of Pamir] (1963) 239; Fl. Kirgiz. 11 (1965) 171; Opred. rast. Sr. Azii [Key to Plants of Mid. Asia] 10 (1993) 555; Gub. Konsp. Fl. Vneshn. Mong. [Conspectus of Flora of Outer Mongolia] (1996) 99. — *A. rupestris* auct. non L.: Poljak. in Fl. SSSR, 26 (1961) 508.

Described from Siberia (Altay). Type in Berlin (B). Map 5.

92 In alpine and subalpine meadows, rocky slopes and talus among stones.

IA. Mongolia: *Khobd.* (Kharkhira mountain range, left bank of Irprak-gol river, 2600 m, among stones, July 16, 1978 — Karam.), *Mong. Alt.* (upper course of Ussein-gol river, alpine tundra, July 27, 1906 — Sap.; upper course of Kharagaitu-gol river on road to Kharagaitu pass, on flank of gorge, July 24; west. Bulgan somon on road to Kharagaitu-khutum, on forest border, July 27–1947, Yun,; Adzhi-Bogdo mountain range, south. flank of Ikhe-gol, sheep's fescue-*Cobresia* slope, 3200 m, Aug. 22, 1979 — Grub., Muld., Dar.; upper Bulgan-gol, gorge of Ulyasutai river, left bank tributary, July 9, 1984 — Kam. et Dar.).

IIA. Junggar: *Tien Shan* (Sairam-nur lake, July 1877; Kash river, June 30, 1879 — A. Reg.; Borotala river basin, before Kokchtau pass, July 1909 — Lipsky; Sairam-nur lake, July 23, 1877 — Fet.).

General distribution: Jung.-Tarb., Nor. and Cent. Tien Shan; East. Sib.

68. A. vulgaris L. Sp. pl. (1753) 848; Franch. Pl. David. 1 (1884) 169; Forbes et Hemsl. Index Fl. Sin. 1 (1888) 446; Krasch. in Kryl. Fl. Zap. Sib. 11 (1949) 2812; Poljak, in Fl. SSSR, 26 (1961) 438; Fl. Kirgiz. 11 (1965) 162; Fl. Kazakhst. 9 (1966) 86; Fl. Sin. 76, 2 (1991) 101; Opred. rast Sr. Azii

[Key to Plants of Mid. Asia] 10 (1993) 549; Gub. Konsp. fl. Vneshn. Mong. [Conspectus of Flora of Outer Mongolia] (1996) 100. – *A. vulgaris* L. var. *xizangensis* Ling et Y.R. Ling in Enum. vasc. pl. Xizang. (1980) 365 and Fl. Xizang. 4 (1985) 770. – *A. vulgaris* L. var. *inundata* Darijmaa in Bull. Soc. natur. Moscou, 5, 97 (1992) 198.

Described from Europe. Type in London (Linn.).

In forest and forest-steppe zones in forest glades, borders, meadows, along river valleys, ravines, among shrubs, around residences, in farms and gardens.

IA. Mongolia: *Depr. Lakes* (in lower courses of Bayantu river, Aug. 15, 1941 – Kondrat'ev), *Val. Lakes* (west. extremity of Begernur lake basin, along gorge, Aug. 23, 1943 – Yun.).

IIA. Junggar: *Cis-Alt.* (10 km north-west of Shara-Sume, in shrubs, July 7, 1959 – Yun.), *Jung. Alat.* (Argaty river, flood-plain forest, Aug. 17, 1957 – Yun.). *Tien Shan* (Sairam-nor lake, July 12, 1878 – Fet.), *Dzhark.* (Kul'dzha, July; same site, Talki gorge, Aug. 1–1877, A. Reg.; same site, Ili river basin, Aug. 7, 1878 – Fet.).

General distribution: Aralo-Casp., Fore Balkh., Jung.-Tarb., Nor. Tien Shan; Arct. Europe, Mediterr., Balk.-Asia Minor, Caucasus, Mid. Asia, West. and East. Sib., Far East, Nor. Mong. (Fore Hubs., Hent., Hang., Mong.-Daur), China (Nor., Nor.-West.), Nor. America.

Note. The polymorphism of this species served as a basis for describing many species and varieties varying essentially in form, size and lanciniation of leaf blade, pubescence, form and sizes of anthodia as well as their arrangement in the panicle. Thus, related species *A. kanashiroi* Kitam. described from East. Mongolia differs from *A. vulgaris* L. in soft-cobwebby pubescence on underside of leaf blade, usually longer than anthodia, and bracts, and orbicular and subobtuse outer phyllary. *A. indica* Willd. is more mesophile with large soft lower cauline leaves, loose broad-pyramidal panicle and sessile, broad-campanulate anthodia. These species are very close to *A. vulgaris* L. and require further study. Because of lack of herbarium material, their geography is cited here after Fl. Intramong. (1993): *A. kanashiroi* Kitam. and *A. indica* Willd. – East. Mongolia (Shilin-Khoto town and Khukh-Khoto).

69. A. xerophytica Krasch. in Not. syst. Herb. Horti Petrop. 3 (1922) 24; Grub. Konsp. fl. MNR [Conspectus of Flora of Mongolian People's Republic] (1955) 270; Leonova in Grub. Opred. rast. Mong. [Key to Plants of Mongolia] (1982) 250; Fl. Intramong. 6 (1982) 155; Fl. Sin. 76, 2 (1991) 19; Fl. Intramong. ed. 2, 4 (1993) 606; Gub. Konsp. fl. Vneshn. Mong. [Conspectus of Flora of Outer Mongolia] (1996) 100. – Ic.: Fl. Intramong. (1982) tab. 55, fig. 9 and 10.

Described from Inner Mongolia (Ordos). Lectotype in St.-Petersburg (LE).

On sandy and rubbly slopes of low mountains, in barren steppes, saxaul thickets, on sandy and clayey-pebble bed banks of rivers and lakes, floors of gorges.

IA. Mongolia: *Khobd.* (42 km north of Bayan-Ulegei, subshrubby feather grass steppe, July 28; 28 km north-west of Bayan-Ulegei centre, July 28 – 1977, Karam.), *Mong. Alt.* (road from Shargain-Gobi to Khan-Taishiri mountain range, wormwood-feather grass steppe, Aug. 24, 1943; 40 km north of Kobdo town, on road to Tsagan-nur, rocky slopes of conical hillock, Aug. 7, 1945 – Yun.; Tsetseg-nur lake basin, barren steppe, Aug. 11, 1945 – Yun.; Adzhi-Bogdo mountain range, Burgastaiin-daba pass, between Indertiin-gol and Ara-Tszuslangiin, Aug. 6–1947, Yun.), *Depr. Lakes* (sand on Dzabkhan river, Sept. 1923 – Pisarev; Shargain-Gobi, west. section, saltwort desert, Aug. 13, 1945 – Yun.; 28 km east of Dzabkhan somon, on road to Ulan-Bator, intermontane plain, feather grass-wormwood steppe on sand, Oct. 5, 1948 – Grub.; 35 km north-west of Urgamal settlement, Sept. 8, 1984 – Gub.), *Val. Lakes* (50 km south of Tugrek settlement, slope of Dzungar-ul town, offshoot of Dzegest-ul, July 28; bank of Tapyn-Tsagan-nur lake, 5 km east of Barun-Bayan-ul settlement, arid steppe, Aug. 2; 45 km north-east of Bayan-tsagan settlement, on sand among almond shrubs, Aug. 4–1983, Gub.), *Gobi Alt.* (Bayan-Tukhum area, hummocky sand, Aug. 29, 1931 – Ik.-Gal.; Ikhe-Bogdo mountain range, on bank of gorge, Oct. 14; Baga-Bogdo mountain range, desert steppe, Sept. 17–1943, Yun.; Arts-Bogdo mountain range, between Dzhiragalantiin-khuduk well and Ikhe-Baga-Ulan, saltwort-feather grass steppe, July 20; nor. trail of Tostu-nuru mountain range, biurgun-nanophyte desert, Aug. 9; Bain-Gobi, Guagan-gol river valley, feather grass steppe, Aug. 27–1948, Grub.; between Bayan-Tsagan and Shine-Dzhinst somons, Aug. 15, 1984 – Kam. et Dar.), *East. Gobi* (north-east of Sain-Shanda, thin sand around lowland, Aug. 29; Bayan-Dzak area, desert steppe, Sept. 1; 50–55 km north of Dalan-Dzadagad, desert, Sept.; Saikhan-dulan mountain range, Barun-Tugrik area, 25 km south-west of Sain-Shanda, Sept. 18; 15 km north-east of Khanygin-Khural, hummocky sand, Sept. 19; Khan-Bogdo, 10 km south-east of somon, sand, Sept. 26–1940, Yun.; 10 km south-east of Undur-Shila somon, desert steppe, June 2; 25 km north-east of Dzamyn-Ude, June 12–1941; 10 km south-east of Elstein-Tobtsog on road to Sain-Shanda, rubbly sand, June 17–1943, Yun.; Bayan-Dzak area, weathered red sandstones, Oct. 21, 1947 – Grub.; Mandal-obo somon, Ulan-nur basin, Aug. 14; 3 km north of Lus somon, steppe with pea shrub, Aug. 15–1950, Lavr.; lower Ongiin-gol river, 3 km north-east of Shine-Usu-khuduk well, Sept. 5, 1950 – Yun.; 15 km south-east of Ulgii settlement, Aug. 4, 1984 – Gub.), *West. Gobi* (10 km west of Dzhinst-ul, thin rubbly sand, June 30, 1941 – Tsatsenkin; 30 km south-east of Atas-Bogdo mountain range 15 km south of Khukhu-usu area, thin sand, Aug. 11–1943; Adzha-Bogdo mountain range, Burgastain-daba pass, Aug. 6–1947, Yun.), *Alash. Gobi* (20 km south of Obotu-Khural, vicinity of Gashun-nur lake and Sogo-nur, Aug. 27, 1949 – Yun.; Bayan-khoto, 25 km west of somon somewhat sand-covered in basin, July 5, 1957 – Petr.; same site, sandy desert, July 5, 1957 – Kabanov; Mintsin, 80 km north of town, vicinity of Baituai-Tengri, sand, July 5; 60 km north of Tszinta station, on arid bed, July 23; Bayan-Khoto, Tengri sand, Aug. 14–1958, Petr.; 120 km south–south-west of Nomgon settlement, Bichigt-Usny-khuduk area and 25 km south-west of Tsagan-Deres border post, July 6; 70 km south-east of Obot-Khural settlement, rubbly desert 10 km east of Tsailyan border post, Aug. 8; 130–140 km south-west of Nomgon settlement, in gorge, Aug. 9–1981, Gub.; "Alashan [Khelanshan]" – Fl. Intramong. (1993) l.c.), *Ordos* ("Bayannor ajmaq [Urot]" – Fl. Intramong. (1993) l.c.), *Khesi* (10 km west of Gaotai town on road to An'chi, Oct. 8, 1957; 23 km south-west of Gaotai town, submontane plain, July 14; same site, 1 km north of An'si, arid bed, July 26–1958; 85 km south-east of Dun'-Khuan somewhat covered by sand, Oct. 7, 1959 – Petr.).

IB. Kashgar: *South.* (60 km south of Nii settlement on road to Yakka settlement, pebble bed, June 8; 67–68 km north of Polur settlement on road to Kerii, knolls, June 13 – 1959, Yun.).

IIA. Junggar: *Jung. Gobi* (20 km north-east of Bodoinchiin-Khure, wormwood-sand desert, Aug. 18, 1947 – Yun.; 9 km from Ubchugiin-gol head on road to Bodongiin-Baishing, thin sand, Sept. 9, 1948 – Grub.; on road to Khami from Guchen, desert steppe along slopes, Oct. 5, 1957 – Yun., Li, Yuan'; 30 km south-west of Bidzh-Altay settlement, rubbly desert, July 25, 1978, Khaldzan-Ul mountain range 50 km south-east of Altay settlement, Sept. 1, 1983 – Gub.).

General distribution: China (Dunbei).

Subgenus 2. Dracunculus (Bess.) Peterm.

Deutsch Fl. (1848) 294; Poljak. in Fl. SSSR, 26 (1961) 525. — *Artemisia* sect. *Dracunculus* Bess. in Bull. Soc. natur. Moscou, 8 (1836) 16. — *Oligosporus* Cass. in Bull. Soc. Philom. Paris (1817) 33; id. in Dict. Sci. Nat. 36 (1826) 25

1. Herbaceous plants with 1–3 or more shoots, sometimes slightly lignifying in lower portion .... 2.

+ Subshrubs and small subshrubs with many, highly (to 5–7–15 cm height) lignifying shoots .... 29.

2. Blades of lower cauline leaves entire, or 3–5-sected at tip or even pinnatisected once into broad-lanceolate terminal leaf lobules .... 3.

+ Blades of lower cauline leaves 2–3-pinnatisected; terminal leaf lobules narrow-lanceolate to filiform .... 13.

3. Leaves of barren shoots and lower cauline leaves long-petiolate .... 4.

+ All leaves sessile .... 6.

4. Leaves of barren shoots and lower cauline leaves on petioles not narrowed toward base and considerably surpassing leaf blade; leaf blade pinnatisected once into short broad-lanceolate lobules short-cuspidate at tip .... 82. A. eriopoda Bge.

+ Leaves of barren shoots and lower cauline leaves on petioles narrowed toward base and not surpassing leaf blade; leaf blade entire dentate along margin or its lobules set at an acute angle .... 5.

5. Lower cauline leaves laciniated into broad-lanceolate lobules; anthodia broad-avoid, short-stalked, nutant; phyllary orbicular at tip .... 77. A. desertorum Spreng.

+ Lower cauline leaves laciniated into narrow-lanceolate or linear lanceolate lobules; anthodia narrow-ovoid, long-stalked, not nutant; phyllary cuspidate at tip ....
....95. A. manshurica (Kom.) Kom.

6. Plant glabrous, green or light brown, sometimes slightly pubescent at commencement of vegetation .... 7.

+ Plant greyish until end of vegetation due to compact tomentose pubescence of short stellate or fairly long forked hairs .... 11.

7. Fertile shoots single or more, (30) 70–150 cm tall, brown or straw-brown, slightly leafy .... 8.

+ Fertile shoots many, 20–30 (40) cm tall, brown or violet-brown, compactly leafy .... 10.

8. Fertile shoots dark brown; leaf blade profile broad-lanceolate; anthodia up to 2 mm long .... 108. A. subdigitata Mattf.

+ Fertile shoots light brown; leaf blade profile lanceolate or linear-lanceolate; anthodia 2.5–3 (4) mm long .... 9.

9. Fertile shoots ribbed; inflorescence, a broad-pyramidal panicle; anthodia 2.5–3 mm long, clustered on lateral branchlets of panicle .... 79. A. dracunculus L.

+ Fertile shoots not ribbed; inflorescence, a narrow-pyramidal panicle; anthodia 3–4 mm long, single or in groups of 2–3 each on lateral branchlets of panicle ....
.... 110. A. waltonii Drumm. ex Pamp.

10. Fertile shoots dark brown or violet-brown; all leaves simple, lanceolate; anthodia 2–3.5 mm long, proximated on lateral branchlets of racemose panicle .... 101. A. pamirica Winkl.

+ Fertile shoots light brown; lower cauline leaves simple, linear-lanceolate, other cauline leaves bipinnatipartite once with 1–2 lobules on both sides; anthodia 5–6 mm long, spaced on lateral branchlets of narrow-pyramidal panicle ....
.... 92. A. kangmarensis Ling et Y.R. Ling.

11 (6). All plants greyish, pubescent with stellate-branched short appressed hairs; cauline leaves entire, linear or lanceolate-linear, with sharply exerted midrib on underside ....
.... 86. A. glauca Pall. ex Willd.

+ Entire plant greyish green, pubescent with long simple or spaced forked hairs; cauline leaves lanceolate to elliptical, dissected at tip into 1–3 (5) teeth .... 12.

12. Fertile shoots many, 30–50 cm tall, ascending at base; leaf blade pubescent on both sides, with 1, rarely 2 teeth at tip ....
.... 73. A. changaica Krasch.

+ Fertile shoots single or more, 50–70 cm tall, erect; leaf blade pubescent only on underside, with 3–5 teeth at tip ....
.... 85. A. girardii Pamp.

13 (2). Annual–biennial plant with slender root; terminal leaf lobules of cauline leaves linear-lanceolate or elongated-ovoid .... 14.

+ Perennial plant with stout root; terminal leaf lobules of cauline leaves narrow-lanceolate or filiform .... 18.

14. Fertile shoots many, greyish, quite compactly pubescent, branched almost from base .... 15.

+ Fertile shoots single or 3–5 together, brown or cervine, glabrous, branched in upper portion .... 17.

15. Lower cauline leaves long-petiolate, petioles enlarged at base, almost semiamplexicaul; leaf blade profile ovoid, bipinnatisected, with 2–3 pairs of lanceolate-linear lobules dentate or entire along margin; cauline leaves sessile .... 16.

96 + Lower cauline leaves short-petiolate, petioles not enlarged at base; leaf blade profile broad-ovoid to suborbicular, 1–2 pinnatisected, with 1–2 pairs of linear or elongated-ovoid lobules entire along margin cauline leaves short-petiolate; anthodia broad-ovoid .... 75. A. demissa Krasch.

16. Terminal leaf lobules dentate along margin; anthodia broad-ovoid, form broad sparse panicle ..... 81. A. edgeworthii Balakr.

+ Terminal leaf lobules entire; anthodia ovoid, form broad compact panicle .... 102. A. pewzowii Winkl.

17. Lower cauline leaves 1–2-pinnatisected; terminal leaf lobules filiform, 8–12 mm long; anthodia broad-ovoid, stalked, nutant, gather on lateral branchlets in racemose inflorescence, forming broad-diffuse panicle .... 72. A. capillaris Thunb.

+ Lower cauline leaves 2–3-pinnatisected; terminal leaf lobules narrow-lanceolate, 3–5 (8) mm long; anthodia ovoid, sessile or on very short stalks, not nutant, gathered on lateral branchlets in spicate inflorescence, forming pyramidal panicle ....
.... 105. A. scoparia Waldst et Kit.

18. Lower cauline leaves and leaves of barren shoots on petioles considerably surpassing oval leaf blade; panicle narrow, racemose .... 19.

+ Lower cauline leaves and leaves of barren shoots on petioles as long as lanceolate or oblong-ovoid leaf blade or shorter; panicle broad-pyramidal or spicate .... 22.

19. Fertile shoots and leaves greyish, compactly pubescent with sericeous hairs; cauline leaves with auricles at base of petiole, bracts pinnatisected once, as long as lateral branchlets of panicles, with anthodia gathered on them in racemose inflorescence .... 20.

+ Fertile shoots light brown, leaves green, glabrous or subglabrous; cauline leaves without auricles at base of petiole, bracts trisected or entire, surpass lateral branchlets of panicle, with anthodia in groups of 2–3 or singly on them ....
.... 97. A. monostachya Bge. ex Maxim.

20. Radical and lower cauline leaves bipinnatisected; terminal leaf lobules lanceolate-linear, cuspidate spinelike at tip; phyllary green .... 103. A. pycnorhiza Ledeb.

+   Radical and lower cauline leaves 1–2 pinnatisected; terminal leaf lobules linear, short-cuspidate at tip; phyllary dark brown .... 21.

21.   Lower cauline leaves on petioles as long as leaf blade; inflorescence, a compact, subspicate panicle .... 98. A nanschanica Krasch.

+   Lower cauline leaves on petioles considerably surpassing leaf blade; inflorescence, a loose pyramidal panicle .... 83. A. frigidioides H.C. Fu et Z.Y. Zhu.

22   Petioles of lower cauline leaves orbicular; terminal leaf lobules linear or narrow-linear, somewhat cuspidate at tip; panicle narrow-pyramidal; anthodia 2–3 (3.5) mm long .... 23.

97   +   Petioles of lower cauline leaves flat; terminal leaf lobules lanceolate, cuspidate spinelike at tip; panicle racemose or spicate; anthodia (3) 4–5 mm long .... 27.

23.   Root rachiform; fertile shoots single or more, brown or slightly erubescent; leaf blade bipinnatisected; terminal leaf lobules narrow-linear, up to 10 (20) mm long .... 25.

+   Rhizome vertical; fertile shoots many, violet-red; leaf blade 1–2-pinnatisected; terminal leaf lobules lanceolate-linear, 5–10 mm long .... 24.

24.   Fertile shoots 8–30 (40) cm tall; lower cauline leaves on petioles considerably surpassing leaf blade; anthodia 2.5–3.5 mm long; phyllary glabrous .... 78. A. dolosa Krasch.

+   Fertile shoots 40–70 cm tall; lower cauline leaves on petioles shorter than leaf blade; anthodia up to 2.5 mm long; phyllary pubescent .... 94. A. macilenta (Maxim.) Krasch.

25.   Fertile shoots herbaceous; leaf blades of lower cauline leaves ovoid in profile; anthodia globose-ovoid, form narrow panicle .... 26.

+   Fertile shoots somewhat lignifying at base; leaf blades of lower cauline leaves broad-oval or orbicular in profile; anthodia ovoid, form broad panicle .... 84. A. gansuensis Ling et Y.R. Ling.

26.   Lower cauline leaves on petioles surpassing leaf blade; terminal leaf lobules linear-lanceolate, 10–30 mm long, cuspidate at tip; anthodia form loose racemose inflorescence on lateral branchlets of panicles; phyllary glabrous .... 74. A. commutata Bess.

+   Lower cauline leaves petioles shorter than leaf blade or as long; terminal leaf lobules lanceolate, up to 10 mm long; anthodia single or in groups of 2–3 on lateral branchlets of panicle; phyllary pubescent .... 100. A. oxycephala Kitag.

27. Fertile shoots lignifying at base and branching from base; panicle pyramidal, with low lateral, 5–10 mm long, branchlets; anthodia hemispherical, sessile; phyllary compactly pubescent .... 80 A. duthreuil-de-rhinsii Krasch.

+ Fertile shoots herbaceous, branching from centre; panicle racemose or spicate, with up to 5 cm long lateral branchlets; anthodia broad-ovoid; phyllary glabrous .... 28.

28. Radical and lower cauline leaves bipinnatisected; anthodia 3–4 mm long. often nutant, form compact, subspicate or compact racemose panicle .... 71. A. borealis L.

+ Radical and lower cauline leaves 1–2-pinnatisected; anthodia 4–5 (6) mm long, erect, form loose racemose panicle .... 76. A. depauperata Krasch.

29 (1). Plant up to 15 cm tall; panicle racemose .... 111. A. wellbyi Hemsl. et Pears.

+ Plant up to 20 cm or more tall; panicle narrow- or broad-pyramidal .... 30.

98 30. Fertile shoots highly branched almost from base, forming subglobose clusters; panicle with short declinate, often interlaced branchlets .... 31.

+ Fertile shoots branched from centre; if branched almost from base, do not form globose clusters; panicle with obliquely erect branchlets .... 33.

31. Entire plant from beginning to end of vegetation compactly pubescent; terminal leaf lobules of lower cauline leaves broad-lanceolate or oblong-ovoid; anthodia narrow-ovoid, sessile .... 32.

+ Entire plant glabrous, greenish brown, toward end of vegetation; terminal leaf lobules of lower cauline leaves linear, 5–12 mm long; anthodia subglobose, stalked .... 106. A. songarica Schrenk.

32. Plant with even more compact, brown pubescence, with somewhat wornout pubescence toward end of vegetation; terminal leaf lobules of lower cauline leaves oblong-ovoid; panicle compact .... 8. A. globosoides Ling et Y.R. Ling.

+ Entire plant greyish, compactly pubescent until end of vegetation; terminal leaf lobules broad-lanceolate; panicle loose .... 87. A. globosa Krasch.

33. Lower cauline leaves and leaves of barren shoots 2–3-pinnatisected; anthodia form compact spicate inflorescence on lateral branchlets of panicle .... 96. A. marschalliana Spreng.

+ Lower cauline leaves and leaves of barren shoots 1–2-pinnatisected; anthodia form compact globose or racemose inflorescence on lateral branchlets of panicle .... 34.

34. Fertile shoots compactly greyish tomentose until end of vegetation; anthodia form compact globose inflorescence on lateral branchlets of panicle .... 109. A. tomentella Trautv.

+ Fertile shoots glabrous toward end of vegetation, from strawbrown to dark brown and almost violet; anthodia single or form loose or compact racemose inflorescence on lateral branchlets of panicle .... 35.

35. Lower cauline leaves and leaves of barren shoots with 1–2 pairs of terminal leaf lobules; anthodia hemispherical or globose; phyllary bulging .... 36.

+ Lower cauline leaves and leaves of barren shoots with 3–5 pairs of terminal leaf lobules; anthodia ovoid; phyllary not bulging .... 38.

36. Lower cauline leaves on petioles as long as blade or shorter; anthodia stalked, nutant, single, in axils of entire and tripartite bracts .... 37.

+ Lower cauline leaves on petioles considerably surpassing blade; anthodia form short, loose racemose inflorescence, in axils of bracts pinnatisected once ....
.... 89. A. gyangzeensis Ling et Y.R. Ling.

37. Petioles of lower and middle cauline leaves not enlarged at base; anthodia 2–3 mm long, form racemose inflorescence on lateral branchlets of panicle; outer phyllary broad-ovoid ....
.... 107. A. sphaerocephala Krasch.

+ Petioles of lower and middle cauline leaves enlarged at base; anthodia 3–5 mm long, form compact globose inflorescence on lateral branchlets of panicles; outer phyllary orbicular ....
.... 112. A. wudanica Liou et W. Wang.

38. Fertile shoots white-waxy in lower portion, brownish yellow above; terminal leaf lobules of cauline leaves 30–40 mm long .....
.... 70. A. albicerata Krasch.

+ Fertile shoots dark brown or almost violet; terminal lobules of cauline leaves 5–30 mm long .... 39.

39. Lower cauline leaves with petioles enlarged at base (semiamplexicaul); anthodia 3–4 mm long .... 40.

+ Lower cauline leaves with petioles not enlarged at base; anthodia 4–5 mm long .... 45.

40. Fertile shoots straw-brown, compactly leafy; panicle narrow-pyramidal; phyllary greenish ....
.... 104. A. saposhnikovii Krasch. ex Poljak.

+ Fertile shoots dark brown or violet, uniformly leafy; panicle pyramidal or broad-pyramidal; phyllary brownish .... 41.

41. Fertile shoots rod-shaped, lignifying for 5–7 cm at base; terminal leaf lobules linear; panicle broad-pyramidal .... 44.

+ Fertile shoots stout, erect, lignifying for 7–15 cm at base; terminal leaf lobules lanceolate-linear; panicle pyramidal ..... 42.

42. Fertile shoots 50–100 cm tall; blades of lower cauline leaves ovoid-lanceolate in profile; terminal leaf lobules 15–30 mm long; anthodia up to 3 mm long .... 43.

+ Fertile shoots up to 50 cm long; blades of lower cauline leaves ovoid or broad-ovoid in profile; terminal leaf lobules 5–15 mm long; anthodia 3–4 mm long ....
.... 115. A. xylorhiza Krasch. ex Filat.

43. Fertile shoots 50–100 cm tall, light brown; blades of lower cauline leaves 3–5 cm long; terminal leaf lobules 15–20 mm long; panicle loose, its lateral branchlets 20–30 cm long ....
.... 99. A. ordosica Krasch.

+ Fertile shoots 30–60 cm tall, dark brown; blades of lower cauline leaves 5–7 cm long; terminal leaf lobules 20–30 mm long; panicle compact, with lateral branchlets up to 15 cm long .... 91. A. intramongolica H.C. Fu.

44. Leaves (except bracts) petiolate, greyish green due to rather diffuse pubescence of long, appressed hairs; phyllary with diffuse pubescence, greenish, subobtuse at tip ....
.... 93. A. klementziae Krasch. ex Leonova.

+ Leaves glabrous, green, lower cauline leaves petiolate, rest sessile; phyllary glabrous, brownish along midrib, cuspidate at tip .... 113. A. xanthochroa Krasch.

45. Fertile shoots dark brown, branched, with branchlets appressed to shoot or subdeclinate; anthodia ovoid, upright, form narrow-pyamidal panicle .... 90. A. halodendron Turcz. ex Bess.

100 + Fertile shoots light brown; anthodia broad-ovoid, nutant, form broad-pyramidal panicle ....
.... 114. A. xigazeensis Ling et Y.R. Ling.

70. A. albicerata Krasch. in Not. syst. (Leningrad), 9 (1946) 173; Poljak. in Fl. SSSR, 26 (1961) 542; Fl. Kazakhst. 9 (1966) 110; Opred. rast. Sr. Azii [Key to Plants of Mid. Asia] 10 (1993) 561. — Ic.: Fl. Kazakhst. 9, Plate 12, fig. 4.

Described from East. Kazakhstan (valley of Ili river). Type in St.-Petersburg (LE).

On hummocky semiconsolidated sand, on sand beds in gorges.

IB. Kashgar: *Nor.* (near Taz-Lyangar village, submontane sand-gravel desert, Aug. 11, 1929 — Pop.; 70 km westward of Karashar settlement, arid river bed somewhat sand-covered, Sept. 10, 1959 — Petr.).

IIA. Junggar: *Jung. Gobi* (Khochin-Elisyn sand, 26 km north of Guchen village, Oct. 3, 1959 — Petr.), *Dzhark.* (right bank of Ili river, 7–8 km south-west of Suidun settlement on road to Santokhodze quay, ridgy semiconsolidated sand, Aug. 31, 1957 — Yun. et Yuan').

General distribution: Fore Balkh., endemic in Cent. Asia.

71. A. borealis Pall. Reise, 3 (1776) 735; Krasch. in Kryl. Fl. Zap. Sib. 11 (1949) 2777; Grub. Konsp. fl. MNR [Conspectus of Flora of Mongolian People's Republic] (1955) 268; Poljak. in Fl. SSSR, 26 (1961) 557; Leonova in Grub. Opred. rast. Mong. [Key to Plants of Mongolia] (1982) 252; Gub. Konsp. fl. Vneshn. Mong. [Conspectus of Flora of Outer Mongolia] (1993) 96. — Ic.: Pall. l.c. Tab. 11, fig. 1; Fl. SSSR, Plate 29, fig. 3.

Described from Siberia (lower course of Ob river). Type in London (BM).

In subalpine and alpine mountain steppes on rocky and rubbly slopes, moraines, coastal meadows along banks of mountain rivers and lakes.

IA. Mongolia: *Khobd.* (22 km south — south-west of Altyn-Tsugts, Togmo-ula mountain, steppe, 2450 m, Aug. 19, 1979. Karam. et al), *Mong. Alt.* (between Dain-gol lake and Ak-Korum, June 29, 1903 — Gr.-Grzh.; upper Kobdo lake, forest meadow, June 27, 1906 — Sap.), *Depr. Lakes* (Dzun-Dzhirgalantu mountain range, south-west. slope of Ulyastyn-gol gorge, 2800 m, June 28, 1971 — Grub., Dar.), *Gobi Alt.* (Gub. l.c.).

General distribution: Arct., West. and East. Sib., Far East, Nor. Mong. (Fore Hubs., Hent., Hang., Mong.-Daur.), Nor. Amer.

72. A. capillaris Thunb. in Nov. Acta Soc. Sci. Upsala, 3 (1780) 209; Forbes et Hemsl. Index Fl. Sin. 1 (1888) 442; Poljak. in Fl. SSSR, 26 (1961) 550; fl. Intramong. 6 (1982) 118; Fl. Sin. 76, 2 (1991) 216; Gub. Konsp. fl. Vneshn. Mong. [Conspectus of Flora of Outer Mongolia] (1996) 96. — *A. scoparia* auct. non Waldst. et Kit.: Fl. Xizang. 4 (1985) 779; Fl. Intramong. ed. 2, 4 (1993) 652. — Ic.: Fl. Intramong. (1982) tab. 39, fig. 10 and 11.

Described from Japan (Pescadores islands). Type in Uppsala (UPS).

On steppe and meadow slopes of low mountains, along gorge floors, rarely on roadsides and on fallow land.

IA. Mongolia: *Cent. Khalkha* (Gub. l.c.), *East. Mong.* (lower Kerulen, before Batszanchi-Sume, 1899—Pot et al.; vicinity of Khailar, sandy steppe, July 18, 1908—Ivashkevich; 375 km east of Urga, Aug. 8, 1927—Ik.-Gal.; Khuntu, 18 km south-east of Bayan-Tsagan somon centre, meadow on ravine floor, Aug. 6, 1946; 12 km west—north-west of Matad somon centre, steppe, Aug. 15, 1949—Yun.; 70 km south-east of Choibalsan town, steppe, Aug. 14, 1954; Bayan-ula mountain, forbs-wild rye meadow, July 8; Tsagan-Obo somon, Enkh-Kholoi area, Aug. 4; 5 km north of Enger-Shavda, slopes of hillocks, Aug. 15-1956, Dashnyam).

General distribution: Far East, Nor. Mong. (Hent., Hang., Mong.-Daur., Fore Hing), China (Dunbei, Nor.), Korean peninsula, Japan.

101

73. A. changaica Krasch. in Acta Inst. Bot. Acad. Sci. URSS, 1, 3 (1936) 346; Grub. Konsp. fl. MNR [Conspectus of Flora of Mongolian People's Republic] (1955) 263; Leonova in Grub. Opred. rast. Mong. [Key to Plants of Mongolia] (1982) 245; Gub. Konsp. Fl. Vneshn. Mong. [Conspectus of Flora of Outer Mongolia] (1996) 97. —Ic.: Grub. Opred rast. Mong. [Key to Plants of Mongolia] (1982), Plate 132, fig. 606.

Described from Mongolia (Hangay). Type in St.-Petersburg (LE).

In steppes, on steppe, rocky and rubbly slopes of low mountains, gorge floors, in ravines and troughs.

IA. Mongolia: *Mong. Alt.* (Gub. l.c.), *Cent. Khalkha* (near Ara-Dzhirgalante river, July 13, 1924—Gorbunova; basin of Ara-Dzhirgalante river, head of Uber-Dzhirgalante river, Aug. 10; same site, between sources of Uber-Dzhirgalante river, and Achit mountain, Aug. 26-1925, Krasch. et Zam.; vicinity of Ikhe-Tukhum-nor lake, Taravdkh mountain, June; same site, Uber-Bumain-ama ravine, July 24-1926, Zam.; vicinity of Undzhul somon, Aug. 11, 1974—Sanczir; same site, Unegtein-Shovgor mountain, in gorge, Aug. 11, 1974—Dar.). *Val. Lakes* (7-8 km from Bayan-Khongor somon, Aug. 28; 40-45 km south—south-east of Arbaikhere centre, in Udazin-nur valley, Aug. 29-1943, Yun.; Naran-Del' river basin, Khutu-Gurt area, Aug. 15; right bank of Tuin-gol river, 15 km above somon, Aug. 16-1949, Kal.).

IIA. Junggar: *Jung. Gobi* (Gub. l.c.).

General distribution: Nor. Mong. (Fore Hubs., Hing.); endemic.

74. A. commutata Bess. in Bull. Soc. natur. Moscou, 8 (1835) 70; Krasch. in Kryl. Fl. Zap. Sib. 11 (1949) 2774; Grub. Konsp. fl. MNR [Conspectus of Flora of Mongolian People's Republic] (1955) 263; Poljak. in Fl. SSSR, 26 (1961) 551; Fl. Kazakhst. 9 (1966) 106; Leonova in Grub. Opred. rast. Mong. [Key to Plants of Mongolia] (1982) 252; Fl. Intramong. 6 (1982) 111; Gub. Konsp. fl. Vneshn. Mong. [Conspectus of Flora of Outer Mongolia] (1996) 96.—*A. campestris* auct. non L.: Franch. Pl. David. 1 (1884) 168; Forbes et Hemsl. Index Fl. Sin. 1 (1888) 442.—*A. pubescens* auct. non Ledeb.: Fl. Intramong. 8 (1985) 327; Fl. Intramong. ed. 2, 4 (1993) 651. —Ic.: Fl. SSSR, Plate 28, fig. 4; Fl. Intramong. (1985) tab. 38, fig. 7-9; Fl. Intramong. (1993) tab. 253, fig. 7, 9.

Described from West. Siberia. Type in Kiev (KW).

On steppe and meadow slopes, borders of larch and birch-larch forests, solonetzic meadows and pebble beds, rarely as weed on fallow land.

IA. Mongolia: *Khobd., Mong. Alat.* (Gub. l.c..), *Cent. Khalkha* (Ikhe-Tukhum-nor, Bayan-Ozura mountain, July 1928 – Zam.), *East. Mong.* (Gurvan-Zagal somon, Shavartyn-obo area, Aug. 20, 1949 – Yun), *East. Gobi* (Gub. l.c.).

General distribution: West. and East. Sib., Far East, Nor. Mong. (Fore Hubs., Hent., Hang., Mong.-Daur., Fore Hing.).

75. A. demissa Krasch. in Acta Inst. Bot. Acad. Sci. USSR, 1, 3 (1936) 348; Enum. vasc. pl. Xizang. (1980) 362; Fl. Intramong. 8 (1985) 329; Fl. Xizang. 4 (1985) 777; Fl. Sin. 76, 2 (1991), 214; Fl. Intramong. ed. 2, 4 (1993) 654; Gub. Konsp. fl. Vneshn. Mong. [Conspectus of Flora of Outer Mongolia] (1996) 97. — *A. implicata* Leonova in Nov. Syst. Pl. Vasc. 17 (1980) 237. — Ic.: Fl. Xizang. (1985) tab. 339, fig. 1–5; Fl. Intramong. (1993) tab. 254, fig. 7, 8.

Described from Qinghai. Type in St.-Petersburg (LE).

On steppe rocky and rubbly slopes of mountains, coastal sand and pebble beds in alpine deserts along arid beds of gorges.

IA. Mongolia: *Mong. Alt.* (Gub. l.c.), *Cent. Khalkha* (vicinity of Ikhe-Tukhum-nor, foothills, Aug. 31, 1926 – Zam.; 5 km east of Choiren-ul mountains, steppe, Aug. 22; 45 km south-east of Choiren, Aug. 25; Khoshun-Shutyin-Gobi area, 2–3 km east of Delger-Tsogtu somon, floor of lowland, Nov. 11–1940, Yun.; vicinity of Sumber somon, Aug. 27, 1970 – Kashapov; 50 km south-east of Nomogun somon, July 30, 1978 – Volk.), *East. Mong.* (vicinity of Dariganga, Baga-Urgo mountain, Aug. 18; same site, Aug. 23; same site, Ikhe-Bulak, aug. 23–1927, Zam.; 40 km north-east of Argalsul-ul mountain, Sept. 9, 1930 – Pob.), *Depr. Lakes* (Gub. l.c.), *Val. Lakes* (south of Tuin-gol river, Sept. 4, 1886 – Pot.; slopes toward Tuin-gol river, Sept. 21, 1924 – Pavl.), *Gobi Alt.* (Tostu mountain range, south of mountain range, Aug. 16, 1886 – Pot.; Dundu-Saikhan mountains, July 13, 1909 – Czet.; Ikhe-Bogdo mountain range, July 10, 1927 – M. Simukova; Dzelin-ula, Sept. 8; same site, west of Bain-Tukhum lake, Sept. 15; Barun-Saikhan mountains, Tegetu-gol area, Sept. 20–1931, Ik.-Gal.; Artsa-Bogdo mountain range, mounds between Dzhirgalant-Khuduk and Ikhe-Bogdo, July 20; rocky ridges between Tsagan-gol river and Ikhe-Bayan-ul, Aug. 1, 1948 – Grub.; Khurkhu mountain range, in saltwort desert, Sept. 8, 1950 – Yun.), *East. Gobi* (Argaly mountain range, vicinity of Khodotyin-khuduk well, Sept. 5, 1928 – Shastin; Del'ger-Khangai, between Lus and Kholtu, Sept. 8–15, 1930 – Kuznetsov; Kalgan road, desert around Dzamyn-Ude, Aug. 19; same site, along slopes of Khukh-Tologain mountains, Aug. 28; same site, 40 km north of Dzamyn-Ude, Motonge mountains, Sept. 30–1931, Pob.; 40 km north-east of Saiin-Shanda somon, Saiin-usu basin, Aug. 23; 16 km north-east of Saiin-Shanda, Tel'ulan-Shaneda area, Aug. 31; 8 km east of Saiin-Shanda, hummocky valley, Sept. 10; 15 km south of Saiin-Shanda, floor of basin, Sept. 14; 30 km west of Saiin-Shanda, Barun-Tugrik area, solonchak lowland, Sept. 17; same site, 1 km west of Ulegei-khid tomb, hummocky sand, Sept. 19; Delgeriin-Deris area, 40 km south-east of Dalan-Dzhirgalan somon, Sept. 24; 10 km north-west of Chandanain somon, Sept. 31–1941, Yun.; Bayan-Dzak-ula area, eroded red sandstones, Sept. 2, 1947 – Grub.; 80 km north-west of Saiin-Shanda, Sept. 5, 1950 – Ivan.; 25 km north-west of Bulgan somon, July 11, 1970 – Banzragch; 75 km north-west of Dkhamyn-ude, debris cone, July 23; 30 km east-south-east of Hubsu-gul somon, July

27–1971, Isach. et Rachk.; 75 km south-east of Mandal-Gobi, June 16, 1972 – Grub. et al.; 80 km south-east of Mandal-Gobi, May 17, 1972; 120 km south – south-west of Saiin-Shanda, on sand, Aug. 3, 1974 – Rachk.; 170 km south-east of Dolan-Dzada-gad town, July 27, 1980 – Gub.), *West. Gobi* (Noyan somon, Khuppu-ula mountain, north-west. trail toward Suchzhi-Khuduk, Aug. 10; same site, saxaul desert, Aug. 17, 1948 – Grub.; on road to Nariin-Bulak from Dzakhoi-Dzaram, saltwort desert, July 6, 1974 – Golubk. et Tsogt.), *Alash. Gobi* (25 km south of Bayan-Khoto, June 10; same site, Alashan desert, sand, Aug. 15, 1958 – Petr.; 40 km north of Obotu settlement, July 25, 1972 – Guricheva et Tsogt; 120 km south-east of Nomgon somon, sandy massif, July 19, 1974 – Rachk.), *Ordos* (15 km east of Dzhasak town, mounds of red sandstone, Aug. 1; 60 km south-west of Kanchin town, on farm fence, Aug. 6; 35 km south-east of Khanchin town, sandy-pebbly plain, Aug. 7, 1957 – Petr.), *Khesi* (23 km south-west of Chzhan'e town, submontane plain, July 17, 1958 – Petr.).

IIA. Junggar: Jung. Gobi (Gub. l.c.).

IIIA. Qinghai: *Nanshan* (valley of Paba-tson river, along Tetung river, July 24, 1908 – Czet.; 33 km west of Xining town, rocky slopes of knolls, steppe, 2950 m, July 1908 – Czet.).

IIIB. Tibet: *South* ("Lhasa, Shigatsze, Tszyantszy" – Fl. Xizang. (1985) l.c.).

General distribution: North Mong. (Hang.), China (Nor., Cent., East.).

76. A. depauperata Krasch. in Animadv. syst. Herb. Univ. Tomsk, 1, 2 (1949) 3; Krasch. in Kryl. Fl. Zap. Sib. 11 (1949) 2777; Grub. Konsp. fl. MNR [Conspectus of Flora of Mongolian People's Republic] (1955) 264; Fl. Kazakhst. 9 (1966) 108; Leonova in Grub. Opred. rast. Mong. [Key to Plants of Mongolia] (1982) 252; Fl. Intramong. ed. 2, 4 (1993) 655; Gub. Konsp. fl. Vneshn. Mong. [Conspectus of Flora of Outer Mongolia] (1996) 97. – *A. pycnorhiza* auct. non Ledeb.: Poljak. in Fl. SSSR, 26 (1961) 560. – Ic.: Fl. Kazakhst. Plate 11, fig. 2; Grub. Opred. rast. Mong. [Key to Plants of Mongolia] Plate 133, fig. 607.

Described from Altay (Chui steppe). Type in St.-Petersburg (LE).

On steppified rocky and rubbly slopes in alpine and subalpine mountain belts, river pebble beds, among rocks.

IA. Mongolia: *Khobd.* (Kharkhira river upper course, on arid slopes, July 21, 1879 – Pot.), *Mong. Alt.* (valley of Urtu-gol river, arid rubbly slope, Aug. 9; Khara-Dzarga mountain range, valley of Sakhir-Sal river, rubbly slopes, Aug. 22–1930, Pob.; Tolbo-Kungei mountain range, alpine steppe, Aug. 5; same site, subalpine steppe, Aug. 5–1945, Yun), *Cent. Khalkha, Depr. Lakes, Val. Lakes* (Gub. l.c.), *Gobi Alt.* (Ikhe-Bogdo mountain range, alpine *Cobresia* areas, Sept. 11, 1943 – Yun.; same site, Uvt gorge, cereal grass-*Cobresia* steppe, June 27, 1972 – Karam. et al), *Alash. Gobi* ("Alashan" – Fl. Intramong. (1993) l.c.).

IIA. Junggar: *Jung. Alt.* (Dzhair mountain range, pass on road to Otu from Toli, mountain steppe, Aug. 9, 1957 – Yun., Li et Yuan').

General distribution: China (West.), Nor. Mong. (Fore Hubs., Hent., Hang., Mong.-Daur.).

77. A. desertorum Spreng. Syst. Veg. 3 (1826) 490; Bess. in Bull. Soc. natur. Moscou, 8 (1835) 64; Pamp. Fl. Carac. (1930) 210; Kitag. Lin. Fl. Mansh. (1939) 428; Poljak. in Fl. SSSR, 26 (1961) 549; Enum. vasc. pl. Xizang. (1980) 362; Leonova in Grub. Opred. rast. Mong. [Key to Plants

of Mongolia] (1982) 252; Fl. Intramong. 6 (1982) 109; Fl. Xizang. 4 (1985) 785; Fl. Sin. 76, 2 (1991) 231; Fl. Intramong. ed. 2, 4 (1993) 660; Gub. Konsp. Fl. Vneshn. Mong. [Conspectus of Flora of Outer Mongolia] (1996) 97.— *A. japonica* auct. non Thunb.: Forbes et Hemsl. Index Fl. Sin. 1 (1888) 443.— *A. desertorum* Spreng. var. *pseudojaponica* Darijmaa et R. Kam. in Bull. Soc. natur. Moscou, 5, 67 (1992) 66. — Ic.: Grub. Opred. rast. Mong. [Key to Plants of Mongolia] Plate 132, fig. 602; Fl. Intramong. (1982) tab. 39, fig. 7–9; Fl. Intramong. (1993) tab. 257, fig. 1–5.

Described from Siberia. Type in Berlin (B).

In steppes, on rubbly slopes of conical hillocks, arid meadows, among shrubs, along river valleys.

IA. Mongolia: *East. Mong.* (In Dzhkhin-tsagan-obo conical hillock region, 9 km northward, July 16, 1970—Grub. et Ulzij.; "Yakeshi, Khailar, Shilingol' ajmaq, Ulantsab, Khukh-Khoto town"—Fl. Intramong. (1993) l.c.).

IIIB. Tibet: *South* ("Lhasa, Nan'mulin"—Fl. Xizang. (1985) l.c.), *Weitzan* ("Bizhu, Sosyan' "—Fl. Xizang. (1985) l.c.).

General distribution: Far East, Nor. Mong. (Hent., Mong.-Daur., Fore Hing.), China (Nor., East.), Korean peninsula.

78. **A. dolosa** Krasch. in Animadv. syst. Herb. Univ. Tomsk. 1, 2 (1949) 2; Krasch. in Kryl. Fl. Zap. Sib. 11 (1949) 2775; Grub. Konsp. fl. MNR [Conspectus of Flora of Mongolian People's Republic] (1955) 264; Leonova in Grub. Opred. rast. Mong. [Key to Plants of Mongolia] (1982) 252; Gub. Konsp. fl. Vneshn. Mong. [Conspectus of Flora of Outer Mongolia] (1996) 97.— *A. commutata* auct. non Bess.: Poljak. in Fl. SSSR, 26 (1961) 552. — Ic.: Grub. Opred. rast. Mong. [Key to Plants of Mongolia] Plate 133, fig. 610.

Described from Siberia (East. Altay). Type in St.-Petersburg (LE).

On steppe rocky, rubbly slopes, pebble beds of mountain rivers, borders of mountain larch forests, rarely on riverine meadows.

IA. Mongolia: *Khobd.* (Gub., l.c.), *Mong. Alt.* (lower course of Turgen'-gol river, left tributary of Bulugun river, Aug. 1947—Yun.), *Cent. Khalkha* (Gub. l.c.), *East. Mong.* (33 km north-west of Khutliin-Khuduk, July 14, 1971—Dashnyam, Karam., Safronova), *Val. Lakes* (Gub. l.c.), *Gobi Alt.* (Ikhe-Bogdo mountain range, south. slope, July 7; Gurban-Saikhan mountain range, east. extremity of Dundu-Saikhan mountain range, rocky slope, July 22–1943, Yun.).

General distribution: West. Sib. (Altay), East. Sib. (Sayans), Nor. Mong. (Fore Hubs., Hent., Mong.-Daur.).

79. **A. dracunculus** L. Sp. pl. (1753) 849; Franch. Pl. David. 1 (1884) 168; Krasch. in Kryl. Fl. Zap. Sib. 11 (1949) 2768; Grub. Konsp. Fl. MNR [Conspectus of Flora of Mongolian People's Republic] (1955) 264; Poljak. in Fl. SSSR, 26 (1961) 529; Fl. Kirgiz. 11 (1965) 176; Fl. Kazakhst. 9 (1966) 105; Leonova in Grub. Opred. rast. Mong. [Key to Plants of Mongolia] (1982) 245; Fl. Intramong. 6 (1982) 113; Fl. Xizang. 4 (1985) 357; Pl. vasc. Helansch. (1986) 357; Fl. Sin. 76, 2 (1991) 187; Fl. Intramong. ed. 2, 4 (1993) 639; Opred. rast. Sr. Azii [Key to Plants of Mid. Asia] 10 (1993)

559; Gub. Konsp. fl. Vneshn. Mong. [Conspectus of Flora of Outer
Mongolia] (1996) 97. —Ic.: Fl. Intramong. (1982) tab. 40, fig. 1–5; Fl.
Intramong. (1993) tab. 249, fig. 1–5.

Described from Siberia. Type in London (Linn.).

On steppe and meadow slopes of mountains, floodplain meadows and
pebble beds, sand, forest borders, glades, in gorges, among shrubs,
standing fallow lands, around residences, roadsides, rarely on solonetzic
meadows, rock talus, in juniper forests.

IA. Mongolia: *Mong. Alt.* (valley of Urtu-gol river, larch forest, Aug. 17; Khara-
Adzarga mountain range, Sakhir-gol river, rocky slope with shrubs, Aug. 22; Khasachty-
Khairkhan mountains, valley of Dundu-Seren-gol river, Sept. 15–1930, Pob.; valley of
Angirta river, Sept. 9; Tsagan-Tyunge area, in lower courses of Uinchi river, sandy
steppe, Sept. 22–1930, Bar.; floodplain of Bulugun river 3–4 km beyond wintering site of
somon, meadow, July 28; Turpon'-gol river, left bank tributary of Bulugun river, arid
meadow, Aug. 7–1947, Yun.), *Cent. Khalkha* (on Ikhe-Tukhum-nor lake—Urga road,
valley of Sasyk-Khuduk river, 60–80 versts (1 verst-1.067 km) from Urga, cereal grass
steppe, Aug. 8, 1925—Krasch. et Zam.; vicinity of Ikhe-Tukhum-nor lake, June 1926—
Zam.; on Ulan-Bator-Choiren-ula road, steppe, July 23, 1941—Yun.; Ulan-Bator-Dalan-
Dzadagad road 72 km south of Ulan-Bator, Aug. 8, 1951—Kal.), *East. Mong.* (Kuku-
Khoto, vicinity of Chaodzhyun'fyn mound, July 18, 1884—Pot.; vicinity of Khorkhonte
station, Oct. 19, 1927—Gordeev; from Choiren to Naran area, Sept. 9; road to Ulan-
Bator-Khoto from Naran area, Ara-Serun mountains, around Dolon urton, Sept. 15–1927,
Zam.; Dariganga, around Archaleul mountains, steppe, Aug. 31; same site, Ongon-Elis
sand, near Boro-Bulak spring, Sept. 13–1931, Pob.; "Yakeshi, Khailar, Khukh-Khoto
town, Baoto"—Fl. Intramong. (1993) l.c.), *Depr. Lakes* (lower course of Buyantu-gol
river, 1942—Kondratenko; 8 km from Ulangom, on gorge floor, July 4, 1977—Karam.,
Sanczir, Sumerina), *Val. Lakes* (Tuin-gol river, on rocky talus, July 9, 1883—Klem.;
valley of Dzabkhyn river, 2 km beyond somon centre, July 8, 1979—Grub., Muld., Dar.),
Gobi Alt. (nor. slope of Tostu mountain range, Aug. 19, 1886—Pot.; Dundu-Saikhan
mountains, on sand bed, July 13, 1909—Czet.; east of Artsa Bogdo, 1925—Chaney; slope
of Dzun-Saikhan mountain, Aug. 25, 1931—Ik.-Gal.; Ikhe-Bogdo mountain range, upper
course of left bank tributaries of Igeru-tag river, Sept. 10, 1943—Yun.; Tostu mountain
range, on nor. slopes, Aug. 15, 1948—Grub.), *West. Gobi* (Tsagan-Bogdo mountain
range, gorge 13 km on road from Tsagan-Bulak spring to Ekhiin-gol, on gorge floor,
Aug. 28, 1979—Grub., Muld., Dar.), *Khesi* (Bei-Shan', Ludzhan-Dzhin' river, Sept. 13,
1890—Gr.-Grzh.).

IB. Kashgar: *West.* (Yarkend-Dar'ya, along irrigation canal, July 6, 1889—Rob.; Ken-
Kol gorge, near Chinese fort, Aug. 11, 1913—Knorr.; Sarykol' mountain range, Bostan-
Terek, July 10, 1929—Pop.).

IIA. Junggar: *Cis-Alt.* (20 km north-west of Shara-Sume, steppe, July 7, 1959—Yun.
et Yuan'), *Tarb.* (on way to Tumanda, Aug. 7, 1876—Pot.; Saur area, on road to Altay
from Karamai, Aug. 4, 1959—Fedorovich), *Jung. Alat.* (Borotala river, Aug. 1878—A.
Reg.; Dzhair mountain range, 1–2 km north of Otu settlement on road to Chugurchak
from Aktam, on bank of brook, Aug. 4, 1959—Yun., Li et Yuan'), *Tien Shan* (Dzhergalant
river, July 29, 1876—Lar.; Kul'dzha, Aug. 17, 1877; Dzhagastai-gol, Aug. 1878—A. Reg.;
Boro-Khoto mountain range, Dzhin-kho area [Tol'o], 1889—Gr.-Grzh.; Tekes river, on
meadow, July 4; Khaidyk-gol river, Sargyn-Bulak area, steppe, Aug. 12–1893, Rob.;
valley of Davanchin river, Sept. 4; near Davanchin village, in fields, Sept. 4–1929, Pop.;
upper course of Tekes river, 7–8 km south-east of Aksu settlement, mountain meadow,
Aug. 24; trails of Ketmen' mountain range, descent into Tekes river valley 7 km east of

Kalmak-Kure, meadow steppe, July 25–1957, Yun., Li et Yuan'; floodplain of Bodonchiin-gol river, 16 km before Bodonchiin-Baishing, in old fields, Sept. 12, 1948 – Grub.; Baitag-Bogdo mountain range, in juniper scrubs, Aug. 8, 1977 – Volk. et Rachk.).

General distribution: Europe, Balk.-Asia Minor, Mid. Asia, Kazakhst., Cent. Tien Shan, West. and East. Siberia, Far East, Nor. Mongolia, China (Dunbei), Nor. Amer.

Note. Highly variable species. *A. dracunculus* L. var. *turkestanica* Krasch. should be differentiated in the mountain steppes of Mongolian Altay and Tien Shan. This variety differs in tall fertile shoots, very short and broad leaf blade and very large anthodia.

80. **A. duthreuil-de-rhinsii** Krasch. in Not. syst. Herb. Horti Petrop. 3 (1922) 22; Enum. vasc. pl. Xizang. (1980) 362; Fl. Xizang. 4 (1985) 786; Fl. Sin. 76, 2 (1991) 234.

Described from Qinghai (Nanshan). Type in St.-Petersburg (LE). Plate V, fig. 2.

In high mountains on clayey and sandy-clayey desertified slopes, rarely on pebble beds of mountain rivers.

IA. Mongolia: *Alash. Gobi* (Alashan mountain range, south. slopes of Yamate gorge, on clayey-sandy soil, May 7, 1908 – Czet.).

IB. Kashgar: *West.* (left bank of Karakash river, 30 km west of Shakhidulla on highway to Kirgiz-Dzhangil pass, pebble bed in valley, June 3, 1959 – Yun. et Yuan').

IIIA. Qinghai: *Nanshan* (nor. slope of Humboldt mountain range, Ulan-bulak area, on clayey slopes, June 24, 1894 – Rob., typus!; same site, Kuitun area, valley of Shara-Gol'dzhin river, sandy-pebbly soil, July 12, 1894 – Rob., paratypus!; same site, Paidza-Tologoi area, Aug. 11, 1894 – Rob., paratypus!).

IIIB. Tibet: *Chang Tang* ("Gaitse" – Fl. Xizang. (1985) l.c.), *South.* ("Lhasa, Shigatsze, Chzhunba, Pulan' " – Fl. Xizang. (1985) l.c.).

General distribution: China (Nor., Nor.-West.).

81. **A. edgeworthii** Balakr. in J. Bombey Nat. Hist. Soc. 63 (1966) 329; Fl. Xizang. 4 (1985) 780; Fl. Sin. 76, 2 (1991) 222. – *A. stricta* Edgew. var. *diffusa* Pamp. in Nouv. Giorn. Bot. Ital. n.s. 34 (1927) 705. – *A. edgeworthii* var. *diffusa* (Pamp.) Ling et Y.R. Ling in Acta Phytotax. Sin. 4, 18 (1980) 509; Enum. vasc. pl. Xizang. (1980) 362. – Ic.: Fl. Xizang. (1985) tab. 339, fig. 8–16.

Described from Tibet. Type in Delhi (DUH).

IIIB. Tibet: *Weitzan* ("Sosyan' " – Fl. Xizang. (1985) l.c.).

General distribution: endemic.

82. **A. eriopoda** Bge. Enum. pl. China bor. (1832) 37; Forbes et Hemsl. Index Fl. Sin. 1 (1888) 442; Franch. Pl. David. 1 (1884) 168; Kitag. Lin. Fl. Mansh. (1939) 425; Fl. Intramong. 6 (1982) 115; Pl. vasc. Helansh. (1986) 357; Fl. Sin. 76, 2 (1991) 235; Fl. Intramong. ed. 2, 4 (1993) 658. – *A. japonica* Thunb. var. *rotundifolia* Debeaux in Acta Soc. Linn. Bot. 33 (1876) 220. – *A. eriopoda* var. *rotundifolia* (Debeaux) Y.R. Ling in Bull. Bot. Res. 8, 4 (1988) 56. – *A. eriopoda* var. *gansuënsis* Ling et Y.R. Ling in Bull. Bot. Res. 8, 3 (1988) 7. – Ic.: Fl. Intramong. (1982) tab. 41, fig. 1–6; Fl. Intramong. (1993) tab. 256, fig. 1–6.

Described from Nor. China. Type in Paris (P).

On steppe slopes, in cereal grass-forbes steppes, 2500-3000 m alt.

IA. Mongolia: *East. Mong.* ("Baotou town, Shilin-Khoto town, Ulantsab, Khukh-Khoto town [Datsin'shan']" — Fl. Intramong. (1993) l.c.), *Alash. Gobi* (Alashan' mountain range, Yamata gorge, June 13, 1908 — Czet.; "Khelan'shan' " — Fl. Intramong. (1993) l.c.).

IIIA. Qinghai: *Nanshan* (Mon'yuan', in Tetungkhe river valley, nor.-east. slope, 2960 m, Aug. 20, 1958 — Dolgushin; 24 km south of Xining town, steppe, 2650 m, Aug. 4, 1959 — Petr.).

General distribution: Far East, China (Dunbei), Korean peninsula, Japan.

106   83. A. frigidioides H.C. Fu et Z.Y. Zhu in Fl. Intramong. 6 (1982) 326, 109. — *A. caespitosa* auct. non Ledeb.: Fl. Intramong. ed. 2, 4 (1993) 616. — Ic.: Fl. Intramong. (1982) tab. 38, fig. 1–6.

Described from Inner Mongolia (East. Gobi-Ulantsab). Type in Hohhot (Khukh-Khoto) (HIMC).

On rocky slopes, pebble beds, along river banks, 2000–3300 m.

IA. Mongolia: *Gobi Alt.* (Dundu-Saikhan mountains, all over mountain range, July 5, 1909 — Czet.), *Alash Gobi* (Alashan mountain range, nor.-west. slopes in Yamate gorge, June 13, 1908 — Czet.).

IIIA. Qinghai: *Nanshan* (Kuku-nor lake, on east. bank, 3210 m, Aug. 5, 1959 — Petr.).

IIIB. Tibet: *Weitzan* (Burkhan-Budda mountain range, nor. rocky slopes in Khatu gorge, July 12, 1901 — Lad.).

General distribution: China (Nor.).

84. A. gansuensis Ling et Y.R. Ling in Bull. Bot. Research, 5, 2 (1985) 9; Fl. Sin. 76, 2 (1991) 214; Fl. Intramong. ed. 2, 4 (1993) 652. — *A. gansuensis* var. *oligantha* Ling et Y.R. Ling in Fl. Intramong. ed. 2, 4 (1993) 651. — Ic.: Fl. Intramong. (1985) tab. 14; Fl. Intramong. (1985) tab. 253, fig. 9, 10.

Described from Qinghai. Type in Beijing (PE).

On loessial plateaus.

IA. Mongolia: *East. Mong.* ("Khukh-Khoto town, Ulantsab, Baotou, Shilingol'sk ajmaq" — Fl. Intramong. (1993) l.c.), *Alash. Gobi* ("Yan'chi, Alashan [Khelan'shan']" — Fl. Intramong. (1993) l.c.), *Ordos* (Bayangol'sk ajmaq — Urat).

General distribution: China (Nor.).

85. A. giraldii Pamp. in Nouv. Giorn. Bot. Ital. n.s. 34 (1927) 657; Pl. vasc. Helansch. (1986) 357; Fl. Sin. 76, 2 (1991) 251; Fl. Intramong. ed. 2, 4 (1993) 662; Gub. Konsp. fl. Vneshn. Mong. [Conspectus of Flora of Outer Mongolia] (1996) 97. — *A. giraldii* var. *longipedunculata* Y.R. Ling in Bull. Bot. Res. 8, 3 (1988) 663. — *A. conaensis* Ling et Y.R. Ling, Fl. Xizang. 4 (1985) 785. — Ic.: Fl. Intramong. (1982) tab. 41, fig. 7–9; Fl. Intramong. (1993) tab. 256, fig. 7–9.

Described from nor.-west. China (Shanxi). Type in Firenze (Florence) (FI).

In mountain steppes, borders of pine forests.

IA. Mongolia: *East. Mong.* ("Ulantsab" — Fl. Intramong. (1993) l.c.), *Alash. Gobi* (Alashan mountain range, Yamota gorge, west. slope, June 5, 1908 — Czet.), *Ordos* (Baga-Edzhin-Khoro area, Aug. 18, 1884 — Pot.).

General distribution: Nor. Mong. (Mong.-Daur.-Gub. l.c.), China (Nor., West.).

86. **A. glauca** Pall. ex Willd. Sp. pl. 3, 3 (1803) 1831; Krasch. in Kryl. Fl. Zap. Sib. 11 (1949) 2770; Grub. Konsp. fl. MNR [Conspectus of Flora of Mongolian People's Republic] (1955) 264; Poljak. in Fl. SSSR, 26 (1961) 535; Fl. Kazakhst. 9 (1966) 106; Leonova in Grub. Opred. rast. Mong. [Key to Plants of Mongolia] (1982) 245; Gub. Konsp. fl. Vneshn. Mong. [Conspectus of Flora of Outer Mongolia] (1996) 97.

Described from Siberia. Type in Berlin (B).

In steppes, on solonetzic, steppified and arid mountain meadows, along banks of rivers, lakes and pebble beds of gorges, in sparse forests and their borders, standing fallow lands, around roadsides and residences.

IA. Mongolia: *Khobd.* (Kharkira mountain range, July 16, 1903 — Gr. — Grzh.; 3–4 km west of Ulan-Daba pass on road to Tsagan-nur, arid mountain steppe, July 29; pass from Uryuknur basin to Achitnur, mountain steppe, July 30; Ulan-Daba pass, Aug. 29 — 1945, Yun.; Duntnur lake, upper belt, Aug. 8–9, 1972 — Metel'tseva), Mong. Altay (Yamatu-gol river, steppe, Oct. 1, 1876 — Pot.), *Depr. Lakes* (Kobdo, July 9, 1870 — Kalning; 20 km south-east of Sagil, Ubsu-nur lake basin, June 30, 1978 — Karam. et al).

IIA. Junggar: *Jung. Gobi* — Gub. l.c.

General distribution: West. and East. Sib., Nor. Mong. (Fore Hubs., Hing., Hang., Mong.-Daur., Fore Hing.), Nor. Amer.

87. **A. globosa** Krasch. in Not. syst. Herb. Horti Petrop. 3 (1922) 27; Grub. Konsp. fl. MNR [Conspectus of Flora of Mongolian People's Republic] (1955) 265; Leonova in Grub. Opred. rast. Mong. [Key to Plants of Mongolia] (1982) 251; Fl. Intramong. 6 (1982) 124; Gub. Konsp. fl. Vneshn. Mong. [Conspectus of Flora of Outer Mongolia] (1996) 97. — Ic.: Fl. Intramong. (1982) tab. 43, fig. 1–6.

Described from Mongolia (Depr. Lakes). Type in St.-Petersburg (LE).

On rocky steppe slopes, borders of mountain forests, arid solonchaks, riverine meadows and pebble beds, in valleys of mountain rivers.

IA. Mongolia: *Khobd.* (20 km south-west of Umne-Gobi somon, Aug. 5, 1979; 16 km north-west of Taralyan somon, Aug. 16, 1980 — Karam. et Beket.), *Mong. Alt.* (Turgen'gol river, left bank tributary of Bulugun river, June 1947 — Yun.), *Cent. Khalkha* (40 km northward of Undurkhan, rocky peak of mound, Aug. 22, 1977 — Dar.), *Depr. Lakes* (Gub. l.c.), *Gobi Alt.* (Ikhe-Bogdo mountain range, south. slope, July 7; Gurban-Saikhan mountain range, lower belt of mountains, Aug. 22 — 1943, Yun.).

IIA. Junggar: *Jung. Gobi* (Khaldzan-ula mountains, on trails, July 30, 1984 — Kam. et Dar.).

General distribution: West. Sib. (Altay), East. Sib. (Sayans).

88. A. globosoides Ling et Y.R. Ling in Bull. Bot. Res. 5, 2 (1985) 7; Fl. Sin. 76, 2 (1991) 199; Fl. Intramong. ed. 2, 4 (1993) 648; Gub. Konsp. fl. Vneshn. Mong. [Conspectus of Flora of Outer Mongolia] (1996) 97. — Ic.: Fl. Intramong. (1993) tab. 253, fig. 1–6.

Described from Mongolia (East. Gobi). Type in Beijing (PE).

On consolidated sand plains.

IA. Mongolia: *East. Mong.* (Shilingol-Khoto town, Ulantsab—Fl. Intramong. l.c.), *East. Gobi* (nor.-east of Bulgan somon, June 20, 1961—Davazhamts; same site, in saxaul forests, on sand, July 31, 1969; 18 km nor. of Bulgan somon, Tevsh-Khairkhan area, an saxaul forests, on sand, Aug. 24, 1970—Sanczir; same site, Sept. 13, 1979—Grub., Muld., Dar.; Bayan-Obo somon, 10 km east of Buduul-Mod post, Aug. 28, 1980—Dar.; Mandal-Obo somon, on consolidated sand plains, Sept. 1, 1976—Sanczir).

General distribution: China (Nor.).

89. A. gyangzeensis Ling et Y.R. Ling in Bull. Bot. Res. 5, 2 (1985) 14; Enum. vasc. pl. Xizang. (1980) 362; Fl. Sin. 76, 2 (1991) 226; Fl. Intramong. ed. 2, 4 (1993) 651. — Ic.: Fl. Xizang. (1980) tab. 510, fig. 8; Fl. Intramong. (1993) tab. 253, fig. 9, 10.

Described from South. Tibet. Type in Beijing (PE).

IA. Mongolia: *East. Mong.* ("Shilingol' ajmaq, Ulantsab, Khukh-Khoto town, Baotou"—Fl. Intramong. (1993) l.c.), *Alash. Gobi* ("Alashan; Helanshan"—Fl. Intramong. (1993) l.c.), *Ordos* ("Bazonor ajmaq, Urat' "—Fl. Intramong. (1993) l.c.).

IIIB. Tibet: *South.* (Gyangze, in declivitate, 3900 m alt., July 12, 1953—Tsoong, typus!).

General distribution: China (East.), Himalayas (Kashmir).

90. A. halodendron Turcz. ex Bess. in Bull. Soc. natur. Moscou, 8 (1835) 17; Grub. Konsp. fl. MNR [Conspectus of Flora of Mongolian People's Republic] (1955) 265; Poljak. in Fl. SSSR, 26 (1961) 539; Leonova in Grub. Opred. rast. Mong. [Key to Plants of Mongolia] (1982) 251; Fl. Intramong. 6 (1982) 124; Fl. Sin. 76, 2 (1991) 191; Fl. Intramong. ed. 2, 4 (1993) 639; Gub. Konsp. fl. Vneshn. Mong. [Conspectus of Flora of Outer Mongolia] (1996) 97. — Ic.: Fl. SSSR, Plate 28, fig. 1; Fl. Intramong. (1982) tab. 42, fig. 13–18; Fl. Intramong. (1993) tab. 250, fig. 13–18.

Described from East. Siberia (Dauria). Type in St.-Petersburg (LE).

On dune, hummocky, often saline sand, sometimes on solonchaks and rubbly slopes of conical hillocks.

IA. Mongolia: *Cent. Khalkha* (vicinity of Ikhe-Tukhum-nor lake, Aug. 27, 1925—Krasch.; Ukhtal somon, hummocky sand on fringe of Shara-Buridu, Sept. 28, 1950—Yun.), *East. Mong.* (Khailar town, Sishan' mounds, July 7, 1951—Skvortsov; same site, Aug. 28, 1951—Li, Tun Pei-jun; Buir-nur lake, Aug. 15, 1977—Dar.; "Yakeshi, Khailar, Manchuria town"—Fl. Intramong. (1993) l.c.).

General distribution: Nor. Mong. (Mong.-Daur., Fore Hing.), China (Dunbei).

91. A. intramongolica H.C. Fu in Fl. Intramong. 6 (1982) 327; Pl. vasc. Helansch. (1986) 357; Fl. Intramong. ed. 2, 4 (1993) 643; Gub. Konsp. fl. Vneshn. Mong. [Conspectus of Flora of Outer Mongolia] (1996) 97. —Ic.: Fl. Intramong. (1982) tab. 44, fig. 1–7; Fl. Intramong. (1993) tab. 251, fig. 1–7.

Described from Inner Mongolia. Type in Khukh-Khoto (HIMC).

On consolidated sand, arid river beds, gorges.

IA. Mongolia: *East. Mong.* (50 km west of Ongon somon, on sand, Aug. 19; 70 km south-west of Bayan-Delger somon, in ravine, Aug. 20–1980, Dar.; west. portion of Shilingol' ajmaq; Shilin-Khoto town—Fl. Intramong. (1993) l.c.; Gub. l.c.), *Gobi Alt.* (Khurmen somon, Tsagan-Kharud-daba pass, in gorge, Sept. 1, 1980—Dar.), *Alash. Gobi* (plain south of Takhiltyn-khyar mountains 21 km south of Noem somon, biurgun desert, Sept. 8, 1979—Grub., Muld., Dar.), *Ordos* (24 km north of Dzhasak town, consolidated sand, Aug. 15, 1957; 55 km west of Dzyutsyuan' settlement, arid bed, Aug. 6, 1958—Petr.).

General distribution: China (Dunbei).

92. A. kangmarensis Ling et Y.R. Ling in Acta Phytotax. Sin. 4, 18 (1980) 510; Enum. vasc. pl. Xizang. (1980) 363; Fl. Xizang. 4 (1985) 789; Fl. Sin. 76, 2 (1991) 195.—*A. salsoloides* auct. non Willd.: Hemsl. Fl. Tibet (1902) p.p., excl. pl. Sib.; Pamp. Fl. Carac. (1930) 209, p.p. excl. pl. Sib.—*A. glauca* auct. non Pall.: Forbes et Hemsl. Index Fl. Sin. 1 (1888) 443, p.p., excl. pl. Sib. et Mong. —Ic.: Ling et Y.R. Ling, Enum. pl. Xizang. (1980) fig. 7.

Described from East. Tibet (Kanma). Type in Beijing (PE).

IIIB. Tibet: *South* (south-west of Lhasa, Kanma village, 4500 m alt., Sept. 15, 1927—Simakova).

General distribution: China (South-West.), Himalayas (west. Kashmir).

93. A. klementziae Krasch. ex Leonova in Novit. syst. pl. vasc. 17 (1980) 236; Grub. Konsp. Fl. MNR [Conspectus of Flora of Mongolian Peoples Republic] (1955) 265; Leonova in Grub. Opred. rast. Mong. [Key to Plants of Mongolia] (1982) 251; Fl. Sin. 76, 2 (1991) 198; Fl. Intramong. ed. 2, 4 (1993) 643; Gub. Konsp. fl. Vneshn. Mong. [Conspectus of Flora of Outer Mongolia] (1996) 97. —Ic.: Grub. Opred. rast. Mong. [Key to Plants of Mongolia] Plate 134, fig. 616.

Described from Mongolia (Hangay). Type in St.-Petersburg (LE).

On semiconsolidated, dune, rather thin and coastal terrace sand.

IA. Mongolia: *Mong. Alt.* (Tsagan-Goliin-Adak steppe, west. Khubchiin-nuru, sand, July 10, 1947—Yun.), *Cent. Khalkha, East. Mong.* (Gub. l.c.), *Depr. Lakes* (between Kurganiin-Turu and Dolon-Turu, sandy steppe, July 21, 1896—Klem.; sand along Dzhabkhyn-gol river, Sept. 1923—Pisarev; 20 km west of Santa-Margats somon, south. bank of Khara-nur lake, hummocky sand, Aug. 18, 1944; Borich-Del sand, south-east of Bayan-nur lake, hummocky sand, July 25, 1945—Yun.; nor. extremity of Barun-Tsakhir mountains 40 km east of Durbel'dzhin somon, June 29, 1980 – Karam.; Bayan-nur lake, Boro-Khoro-Elesu sand, Aug. 15, 1972 – Metel'tseva), *Gobi Alt.* (Arts-Bogdo mountain range, 4 km west of Dzhirgalain-Khuduk, steppe, Aug. 28, 1948 – Grub.; Dzhinst-ula mountains, Aug. 5, 1984 – Karam. et Dar.).

General distribution: Nor. Mong. (Hang., Mong.-Daur.); endemic.

94. **A. macilenta** (Maxim.) Krasch. in Mat. po fl. i rast. SSSR, 2 (1946) 156; Poljak. in Fl. SSSR, 26 (1961) 550; Fl. Intramong. 6 (1982) 120; Fl. Intramong. ed. 2, 4 (1993) 651. – *A. campestris* L. var. *macilenta* Maxim. in Mem. Ac. Sci. Petersb. Div. Sav. 9 (1859) 158 [Prim. Fl. Amur]. – Ic.: Fl. Intramong. (1982) tab. 38, fig. 10, 11; Fl. Intramong. (1993) tab. 254, fig. 9, 10.

Described from Far East. Type in St.-Petersburg (LE).

In steppes on slopes of low mountains, forest borders, among shrubs, in river valleys.

IA. Mongolia: *East. Mong.* (50 km north-west of Ongon somon centre, Aug. 19; west of Bayan-nur lake, Aug. 19–1980, Dar.; "Khukh-Khoto town [Datsinshan']" – Fl. Intramong. (1993) l.c.).

General distribution: Far East, China (Dunbei), Nor. Mong. (Hang., Mong.-Daur.).

95. **A. manshurica** (Kom.) Kom. in Komarov and Klobukova-Alisova, Opred. rast. Dal'nevost. kraya [Key to Plants of Far Eastern Region] 2 (1932) 1035; Fl. Sin. 76, 2 (1991) 240; Fl. Intramong. ed. 2, 4 (1993) 657; Gub. Konsp. fl. Vneshn. Mong. [Conspectus of Flora of Outer Mongolia] (1996) 98. – *A. japonica* Thunb. var. *manshurica* Kom. in Acta Horti Petrop. 25, 2 (1907) 656 [Fl. Mansh. 3, 2]; Kitag. Lin. Fl. Mansh. (1939) 427; Fl. Intramong. 6 (1982) 112. – Ic.: Fl. Intramong. (1982) tab. 41, fig. 11-13; Fl. Intramong. (1993) tab. 256, fig. 10-13.

Described from Manchuria. Type in St.-Petersburg (LE). Plate V, fig. 1. Map 7.

On steppe and meadow slopes, in feather grass-cereal grass and forbs-cereal grass steppes, on arid meadows, fallow land, rarely on sandy soils of submontane plains.

IA. Mongolia: *Cent. Khalkha*-Gub. l.c., *East. Mong.* (Dalainor, 1870 – Lom.; Kerulen, near Berlik mountains, 1899 – Pal.; Khailar, July 6, 1901 – Menzhinskii; 44 km west-north-west of Choibalsan, Aug. 1, 1949 – Yun.; 3 km south-east of Khalkha-gol somon on road to Bayan-Tsagan, Aug. 5; 5 km from Khuntu somon, upper portion of conical hill, Aug. 7; 24-30 km south-east of Khuntu somon, Aug. 8; 12 km west – north-west of Matad somon, Aug. 15; 10 km north-west of Zodal-Khan-ula somon, steppe, Sept. 13 – 1949, Yun.; vicinity of Khailar town, meadow on sandy plain, Aug. 29, 1951 – A.R. Lee (1959); "Shilin-Khoto" – Fl. Intramong. (1993) l.c.).

General distribution: Far East, North Mong. (Hent., Hang., Mong.-Daur.), China (Dunbai).

96. **A. marschalliana** Spreng. Syst. Veg. 3 (1826) 496; Krasch. in Kryl. Fl. Zap. Sib. 11 (1949) 2773; Fl. Kazakhst. 9 (1966) 107; Gub. Konsp. fl.

Vneshn. Mong. [Conspectus of Flora of Outer Mongolia] (1996) 98. — *A. inodora* M.B. Fl. taur.-cauc. 2 (1808) 295. — *A. campestris* auct. non L.: Poljak. in Fl. SSSR, 26 (2961) 553, p.p.; Opred. rast. Sr. Azii [Key to Plants of Mid. Asia] 10 (1993) 562. — Ic.: Fl. Kazakhst. Plate 11, fig. 4.

Described from Crimea and Caucasus. Type lost.

In arid shrubby steppes, steppe meadows, sand, steppe pine forests, rarely on rocky and rubbly slopes of low mountains, in river valleys.

110　　IA. Mongolia: *Mong. Alt.* (Gub. l.c.).

IIA. Junggar: *Cis-Alt.* (20 km north-west of Shara-Sume on road to Kran, Shrubby steppe, July 7, 1959 — Yun.), *Tarb.* (east. trail of Saur mountain range, on road to Burchum from Kosh-Tologoi, steppe, July 4, 1959 — Fedorovich), *Jung. Gobi* (valley of Urumchi river, 3-5 km southward, chee grass thickets on terrace, July 9, 1959 — Yun.).

General distribution: Aralo-Casp., Fore Balkh., Jung.-Tarb., West. Sib.

Note. Distinct species differing from *A. campestris* L. in several morphological characteristics; shorter and narrower terminal leaf lobules and narrow-ovoid anthodia gathered spicately on lateral branchlets of narrow pyramidal panicle. Zonal species of arid shrubby steppes.

97. **A. monostachya** Bge. ex Maxim. in Bull. Ac. Sci. Petersb. 17 (1872) 429; in clave: Grub. Konsp. fl. MNR [Conspectus of Flora of Mongolian People's Republic] (1955) 267; Leonova in Grub. Opred. rast. Mong. [Key to Plants of Mongolia] (1982) 252; Fl. Intramong. 8 (1985) 325; Gub. Konsp. fl. Vneshn. Mong. [Conspectus of Flora of Outer Mongolia] (1996) 96. — *A. bargusinensis* auct. non Bge. ex Maxim.: Fl. Intramong. ed. 2, 4 (1993) 655.

Described from Mongolia. Type in Paris (P). Isotype in St.-Petersburg (LE).

In steppes on rocky and rubbly slopes of low mountains and conical hillocks, on banks of pebble beds, riverine terraces.

IA. Mongolia: *Khobd.* (30 km north — north-east of Umne-Gobi somon centre, steppe, July 21, 1978 — Karam.), *Mong. Alt.* (Kobdo river basin, east. bank of Durge-nur lake, steppe, June 30, 1971 — Grub.), *Depr. Lakes* (10 km east of Airik-nur, July 20, 1945 — Yun.; 37 km north-east of Kobdo river, steppe, Aug. 1, 1977 — Karam., Sanczir, Sumerina), *East. Mong., Gobi Alt.* (Gub. l.c.).

IIA. Junggar: *Jung. Gobi* (Gub. l.c.).

General distribution: East. Sib. (Sayans), Far East, Nor. Mong. (Fore Hubs., Hent., Hang., Mong.-Daur.), China.

98. **A. nanschanica** Krasch. in Not. syst. Herb. Horti Petrop. 3 (1922) 19; Enum. vasc. pl. Xizang. (1980) 363; Fl. Xizang. 4 (1985) 781; Fl. Sin. 76, 2 (1991) 245. — Ic.: Fl. Xizang. (1985) tab. 341, fig. 1–10.

Described from Qinghai (Nanshan). Type in St.-Petersburg (LE).

In alpine belt of mountains on clayey-rocky slopes, meadows, rocks.

IIIA. Qinghai: *Nanshan* (Nanshan mountain range, alps, June 11, 1879 — Przew., typus!; Humboldt mountain range, alpine meadow, June 30, 1894 — Rob.; Kuku-nor lake, meadow on east. bank, 3210 m, Aug. 5, 1959 — Petr.).

IIIB. Tibet: *Chang Tang* (Przewalsky mountain range, nor. slope, on rocks, Aug. 23, 1890 — Rob.; Taitsze, Shuankhu, Ban'ge, An'do, Zhetu" — Fl. Xizang. (1985) l.c.), *Weitzan* (Russkoe lake, nor.-west. bank of Dzhagyn-gol river, July 8–26; up along Tala-chyu river, July 14–26–1884, Przew.; Russkoe lake, July 23, 1900; nor. slope of Burkhan-Budda mountain range, Khatu gorge, June 12–1901, Lad.).

General distribution: endemic.

99. A. ordosica Krasch. in Not. syst. (Leningrad), 9 (1946) 173; Fl. Intramong. 6 (1982) 124; Pl. vasc. Helansch. (1986) 357; Drevesn. rast. Tsinkhaya [Wooded Plants of Qinghai] (1987) 635; Fl. Intramong. (1993) tab. 252, fig. 1–6; Fl. Sin. 76, 2 (1991) 195; Fl. Intramong. ed. 2, 4 (1993) 647; Gub. Konsp. Fl. Vneshn. Mong. [Conspectus of Flora of Outer Mongolia] (1996) 98. — Ic.: Fl. Intramong. (1982) tab. 46, fig. 1–6.

Described from Mongolia (Ordos). Type in St.-Petersburg (LE). Plate V, fig. 5. Map 7.

On semiconsolidated, dune sand, sand-covered arid beds and pebble beds.

IA. Mongolia: *Mong. Alt.* (Gub. l.c.), *East. Mong.* ("Ulantsab, Khukh-Khoto town" — Fl. Intramong. (1993) l.c.; Gub. l.c.), *Gobi Alt.* (east. extremity of Tosten-nuru mountain range, 8 km north-east of Demin-Khuduk well, Sept. 4; near Elstein-Dzagdai spring, Sept. 6; 4 km west of Noën-Bogda mountains, in gorge, Sept. 8; nor. extremity of Noën-Nuru mountains, sand, Sept. 8; Bayan-Tukhum lake, consolidated sand, Sept. 9–1979, Grub., Muld., Dar.), *East. Gobi* (22 km south of Hubsugul somon, desert steppe, Aug. 22; 80 km south of Hubsugul somon, Aug. 23–1980, Dar. et Zumbelma), *West. Gobi* (16 km north-west of Shinginst settlement, on road to Kushui, along gorge, Sept. 6; 30 km east of An'si, south. extremity of Beishan, Oct. 7–1959, Yun.; 35 km south of Shine-Zhinst somon, Aug. 8, 1979 — Sanczir), *Alash. Gobi* (road to Urgu from Alashan, Shara-Burdu area, May 5, 1909 — Czet.; vicinity of Bayan-Khoto mountains, dune sand, July 5, 1957 — Kabanov; 15 km south-east of Min'tsin town, sand-covered solonchak meadow, July 3; Bayan-Khoto sand, July 30–1958; Min'tsin, west. extremity of oasis, somewhat sand-covered, Aug. 11; 36 km east of Min'tsin, semiconsolidated sand, Aug. 18–1959, Petr.), *Ordos* (Huang He river valley, Aug. 24, 1871 — Przew., typus!; same site, near Termin-bashin village, Aug. 10; Ulan-Morin area, Aug. 23–1884, Pot.; Bayanur ajmaq, Urot, Khodtsy, Donnou — Fl. Intramong. (1993) l.c.), *Khesi* (50 km east of An'si settlement, July 24; 36 km north of An'si settlement, arid sand-pebble bed, July 27–1958, Petr.).

IC. Qaidam: *Plain* (20 km north-west of Chichin's settlement, in arid river bed, Oct. 5; 8 km westward of Yunchan settlement, Oct. 8; 14 km north of Tsagan-us, somewhat sand-covered, Oct. 14 — Petr.).

General distribution: endemic in Cent. Asia.

100. A. oxycephala Kitag. in Rep. First Sci. Exped. Mansh. 4, 4 (1936) 93; Fl. Intramong. 8 (1985) 327; Fl. Sin. 76, 2 (1991) 200; Fl. Intramong. ed. 2, 4 (1993) 647; Gub. Konsp. fl. Vneshn. Mong. [Conspectus of Flora of Outer Mongolia] (1996) 98. — *A. pubescens* Ledeb. var. *oxycephala* (Kitag.) Kitag. Lin. Fl. Mansh. (1939) 429. — Ic.: Fl. Intramong. (1985) tab. 43, fig. 9, 10; Fl. Intramong. (1993) tab. 252, fig. 9, 10.

Described from Manchuria. Type in Tokyo (TI).

In sandy steppes.

IA. Mongolia: *Cent. Khalkha* (Gub. l.c.), *East. Mong.* (between Bayan-nur area and Eren-tsab, steppe, Aug. 2, 1985 — Kam. et Dar.; "Yakeshi" — Fl. Intramong. (1993) l.c.).

General distribution: Nor. Mong. (Mong.-Daur., Fore Hing.), China (Dunbei).

101. A. pamirica Winkl. in Acta Horti Petersb. 11, 2 (1872) 329; Grub. Konsp. Fl. MNR [Conspectus of Flora of Mongolian People's Republic] (1955) 287; Poljak. in Fl. SSSR, 26 (1961) 534; Ikonn. Opred. rast. Pamira [Key to Plants of Pamir] (1963) 239; Fl. Kirgiz. 11 (1965) 179; Fl. Kazakhst. 9 (1966) 106; Enum. vasc. pl. Xizang. (1980) 363; Leonova in Grub. Opred. rast. Mong. [Key to Plants of Mongolia] (1982) 245; Fl. Xizang. 4 (1985) 782; Opred. rast. Sr. Azii [Key to Plants of Mid. Asia] 10 (1993) 559; Gub. Konsp. Fl. Vneshn. Mong. [Conspectus of Flora of Outer Mongolia] (1996) 98. — *A. dracunculus* auct. non L.: Pamp. Fl. Carac. (1930) 209.

Described from Pamir (Kara-Kul' lake). Type in St.-Petersburg (LE).

On rocky, rubbly and melkozem slopes of alpine and subalpine mountain belts, in valleys of mountain rivers, in moraines.

IA. Mongolia: *Khobd.* (pass from Uryuk-nur lake basin to Atat-nur lake, July 3; 10 km along Khoiligin-Daba from Tsagan-nur somon on road to Bayan-Ulegei, mountain steppe, Aug. — 1945, Yun.), *Mong. Alt.* (Taishiri-Ola mountain range, July 15, 1877 — Pot.; Adzhi-Bogdo mountain range, Dzusylyn gorge, July 29, 1877 — Pot.; near Bol'shoi Ulan-Daba pass, July 19, 1898 — Klem.; upper course of Uinta river, Sept. 7; on way to salt lake from Angyrta, Sept. 9, 1930 — Bar.; Tolbo-Kungei mountain range, arid steppe, Aug. 5, 1945; Adzhi-Bogdo mountain range, upper course of Mainigti-amo creek valley, Aug. 1947 — Yun.; valley of Khobdo river on Khobdo-Ulangom road, meadow, Aug. 23, 1944 — Yun.), *Val. Lakes* (right bank of Ongiin-gol river, Aug. 25, 1893 — Klem.; Maidakhtsyn-ama area, between Tsagan-Olom and Dzag-Baidarik, steppe, Aug. 27, 1943 — Yun.), *Gobi Alt.* (Mailarun-ama area, Aug. 27, 1945 — Yun.; Gurban-Saikhan mountain range, Aug. 9, 1970 — Sanczir; Nemegetu-ula, on south. slope, July 29, 1972 — Guricheva; Gurban-Saikhan mountain range, Aug. 4, 1972 — Isach.).

IB. Kashgar: *West.* (Turer area, July 27, 1908 — Divn.; Sulu-Sakhal river, 25 km east of Irkeshtam, 2800-2900 m, July 26, 1935 — Olsuf'ev).

IIIA. Qinghai: *Nanshan* (Yuzhno Kukunor mountain range, Tangut area, nor. slope, 2800 m, June 23, 1880 — Przew.; same site, Usubin-gol river, 3300 m, Aug. 16, 1901 — Lad.), *Amdo* (Huang He river, near Guidui town, 2500 m, June 23, 1880 — Przew.).

IIIB. Tibet: *Chang Tang* (nor. slope of Russky mountain range, on river pebbles, June 2, 1890 — Rob.).

IIIC. Pamir (Pas-Rabat, July 3, 1909 — Divn.; midcourse of Kanlyk river, 2500-3200 m, July 12; upper course of Lanet river, on moraine, July 20-1942, Serp.).

General distribution: Jung.-Tarb., Nor. and Cent. Tien Shan, East. Pamir, endemic in Cent. Asia.

102. A. pewzowii Winkl. in Acta Horti Petrop. 13, 1 (1893) 3; Enum. pl. vasc. Xizang. (1980) 364; Fl. Xizang. 4 (1985) 780; Fl. Sin. 76, 2 (1991) 225; Opred. rast. Sr. Azii [Key to Plants of Mid. Asia] 10 (1993) 563. — *A. demissa* auct. non Krasch.: Poljak. in Fl. SSSR, 26 (1961) 562; Ikonn. Opred. fl. Pamira [Key to Plants of Pamir] (1963) 240.

Described from Tibet (Bostan-Tograk river). Type in St.-Petersburg (LE). Plate VI, fig. 4.

On loessial, clayey and rubbly slopes.

IB. Kashgar: *South.* (Keriya, July; same site, Chivei area, July 3–1885, Przew.; 10 km north of Polur on road to Keriya, 3000 m, sand-covered rubbly slopes, May 3; nor. slope of Kuen'-Lun' mountain range, 60 km south-east on road to Polur, 2200 m, loessial slopes, May 10–1959, Yun. et Yuan).

IIIB. Tibet: *Chang Tang* (Russky mountain range, nor. slope of Bostan-Tograk river, loessial slopes, July 11, 1880 – Rob., typus!).

General distribution: Mid. Asia (Pam.-Alay).

103. A. pycnorhiza Ledeb. in Fl. Alt. 4 (1833) 79; Krasch. in Kryl. Fl. Zap. Sib. 11 (1949) 2775; Grub. Konsp. fl. MNR [Conspectus of Flora of Mongolian People's Republic] (1955) 267; Poljak. in Fl. SSSR, 26 (1961) 559, excl. syn. *A. depauperata* Krasch.; Leonova in Grub. Opred. rast. Mong. [Key to Plants of Mongolia] (1982) 252; Fl. Intramong. ed. 2, 4 (1993) 655; Opred. rast. Sr. Azii [Key to Plants of Mid. Asia] 10 (1993) 562, excl. syn. *A. depauperata* Krasch.; Gub. Konsp. fl. Vneshn. Mong. [Conspectus of Flora of Outer Mongolia] (1996) 99.

Described from Altay (Chuya steppe). Type in St.-Petersburg (LE).

In mountain steppes on rocky, rubbly slopes, limestone outcrops, on pebble beds of mountain rivers, in rock crevices.

IA. Mongolia: *Khobd.* (west – north-west of Ulangom, Shuru-Ulyasty-gol river gorge, steppe, July 5, 1977 – Karam., Sanczir, Sumerina; same site, 50 km west of Ulangom, Morto-ula massif, June 20, 1978 – Karam. et Beket), *Mong. Alt.* (Dolon-nur, on granites, July 8, 1877 – Pot.; Taishir-ula mountain range, in rocks, Aug. 15, 1930–Pob.; nor. slope of Mong. Altay, Khalyun area, rocky slopes, Aug. 24, 1943; 25–27 km east of Tsetseg-nur, on road to Tonkhil' somon, steppe, Aug. 12, 1945; 10 km east of Yusun-bulak, steppe, July 14; upper course of Bulugun river, tributary of Indertiin-gol, steppe, Aug. 25–1947, Yun.; Khan-Taishiri mountain range, Khabchingin-daba pass, on Yusun-Bulak – Tonkhil' somon road, steppe, Sept. 31, 1948 – Grub.), *Cent. Khalkha* (Choiren-ula mountain, in rock crevices, July 7, 1941 – Yun.), *Depr. Lakes* (Khan-Khukhei mountain range, south-west. slope of Dulan-ul, July 21, 1945 – Yun.), *Val. Lakes* (right bank of Tuin-gol river, steppe, June 29, 1941 – Tsatsenkin), *Gobi Alt.* (Khalga pass, between Barun-Saikhan and Dundu-Saikhan, on rubbly slope, Aug. 3, 1931 – Ik.-Gal.; pass between Dzun and Dundu-Saikhan, along floor of arid ravine, July 22; Ikhe-Bogdo mountain range, steppe, Sept. 14; same site, arid mountain steppe, Sept. 14; same site, cereal grass steppe on nor. slope of Bogdo-ul mountain, Sept. 14; steppe on rocky slope, Sept. 15–1943, Yun.; same site, Narin-khurilt gorge, July 30; west. extremity of Bayan-Tsagan-ul, 21 km from Bayan-Tsagan somon, on limestone outcrop, Aug. 28, 1948 – Grub.; same site, Ulyasutai creek valley, Aug. 1, 1972 – Banzragch, Karam., Sumerina; Arts-Bogdo mountain range, 30 km from Bogdo-Uvei-Khangai settlement, along gorge floor, 2000 m, July 30, 1983 – Gub.), *Alash. Gobi* (" *Luntoushan*" – Fl. Intramong. (1993) l.c.).

General distribution: West. Sib. (Altay), Nor. Mong. (Fore Hubs., Hent., Hang., Mong.-Daur.).

104. A. saposhnikovii Krasch. ex Poljak. in Not. Syst. (Leningrad), 17 (1955) 412; Poljak. in Fl. SSSR, 26 (1961) 542; Fl. Kirgiz. 11 (1965) 176; Fl. Sin. 76, 2 (1991) 203; Opred. rast. Sr. Azii [Key to Plants of Mid. Asia] 10 (1993) 561. — Ic.: Fl. SSSR, 26, Plate 30, fig. 3; Fl. Kirgiz. Plate 8, fig. 2.

Described from Cent. Tien Shan (Imyl'chek river). Type in St.-Petersburg (LE). Map 7.

On rocky slopes of mountains, pebble beds of mountain rivers, in gorges, among boulders.

IB. Kashgar: *Nor.* (east. part of Baisk depression, on road to Kyzyl settlement, desert along gorges, Sept. 1; on road from Kyzyl settlement of Aksu, desert along gorges, Sept. 3–1958, Yun.).

IIA. Junggar: *Tien Shan* (valley of Muzart river, 7–8 km above Oi-Terek area on road to Aksu, chee grass thickets along gorge floor, Sept. 8; right summit of Tanu-Davan near Oi-Terek area, Sept. 9; 60 km north-west of Aksu settlement, Sept. 17–1958, Yun.).

General distribution: Cent. Tien Shan; endemic in Cent. Asia.

105. A. scoparia Waldst. et Kit. Icon. Pl. Rar. Hung. 1 (1801) 66; Franch. Pl. David, 1 (1884) 167; Forbes et Hemsl. Index Fl. Sin. 1 (1888) 445; Kitag. Lin. Fl. Mansh. (1939) 431; Krasch. in Kryl. Fl. Zap. Sib. 11 (1949) 2778; Grub. Konsp. Fl. MNR [Conspectus of Flora of Mongolian People's Republic] (1955) 269; Poljak. in Fl. SSSR, 26 (1961) 560; Fl. Kirgiz. 11 (1965) 179; Fl. Kazakhst. 9 (1966) 114; Enum. vasc. Pl. Xizang. (1980) 364; Leonova in Grub. Opred. rast. Mong. [Key to Plants of Mongolia] (1982) 246; Fl. Intramong. 6 (1982) 115; Fl. Xizang. 4 (1985) 779, excl. syn. *A. capillaris* Thunb.; Pl. vasc. Helansch. (1986) 357; Fl. Sin. 76, 2 (1991) 220; Fl. Intramong. ed. 2, 4 (1993) 652; Opred. rast. Sr. Azii [Key to Plants of Mid. Asia] 10 (1993) 563; Gub. Konsp. fl. Vneshn. Mong. [Conspectus of Flora of Outer Mongolia] (1996) 99. — Ic.: Waldst. et Kit. l.c. (1801) tab. 65; Fl. Intramong. (1982) tab. 39, fig. 1-6; Fl. Xizang. (1985) tab. 340, fig. 1–5; Fl. Intramong. (1993) tab. 264, fig. 1–6.

Described from Europe (Hungary). Type in Wien (Vienna) (W).

On rubbly, rocky and melkozem slopes and trails of mountains in steppes and deserts, solonetzic meadows, sand, loamy sand, as weed in pastures, fallow land and farms, coastal pebble beds, in gorges and chee grass thickets.

IA. Mongolia: *Mong. Alt.* (upper course of Bodonchiil-gol, September 8, 1930 — Bar.; upper course of Bulugun-gol, at confluence of Ulyasutain-gol river into it, meadow, July 20, 1947 — Yun.; 34 km north-east of Altay somon, Kobdo river, July 2, 1979 — Karam.), *Cent. Khalkha* (midcourse of Ubur-Dzhirgalante river, Aug. 29, 1925 — Krasch. et Zam.; 4 km south-east of Undzhul somon centre, Aug. 24, 1974 — Dar.), *Depr. Lakes* (Shargiin-Gobi, intersection of Gobi desert from Tanshir mountain range toward Khalgon river, Aug. 16, 1930 — Pob.; lower course of Buyantu-gol river, July–Aug. 1941 — Kondrat'ev, 10 km north-west of Sagil somon centre, July 30, 1978 — Karam.; 30 km south-east of Khara-Us-nur lake, submontane rocky desert, July 24, 1979 — Gub.; 32 km south-east of

Dzabkhan somon, forbs-cereal grass steppe, July 2; same site, 34 km from somon, in ravine, Aug. 4–1980, Karam.).

IB. Kashgar: *Nor.* (valley of Muzart river, Aug. 1, 1877 – Fet.; Chubogoin-nor area, along irrigation ditch, Aug. 17, 1893 – Rob.).

IIA. Junggar: *Cis-Alt.* (15 km north of Barbagai, on mountain slope, July 14, 1959 – A.R. Lee (1959), *Jung. Alt.* (25 km north of Toli settlement, along roadsides, Aug. 9, 1957; 4 km south-west of Dubukhe, Sept. 1, 1957, No. 3163 – Kuan), *Tien Shan* (valley of Ili river at inflow of Tekes river into it, June 29, 1876 – Przew.; Boro-Khoro mountain range, Dzhagastai gorge, Aug. 1878 – A. Reg.; Ketmen' mountain range, from Kul'dzha to Kzyl-Kure, intermontane valley, Aug. 23, 1957 – Yun., Li et Yuan'; vicinity of Bulgan somon, desert, Aug. 2, 1978 – Volk.).

IIIB. Tibet: *South.* ("Lhasa, Shigatsze" – Fl. Xizang. l.c.).

General distribution: Aralo-Casp., Fore Balkh., Jung.-Tarb., Nor. Tien Shan, Europe, Balk.-Asia Minor, Caucasus, Mid. Asia, West. and East. Sib., Nor. Mong. (Hent., Hang., Mong.-Daur., Fore Hing.).

106. A. songarica Schrenk in Fisch. et Mey., Enum. pl. nov. 1 (1841) 49; Poljak. in Fl. SSSR, 26 (1961) 543; Fl. Kazakhst. 9 (1966) 113; Fl. Sin. 76, 2 (1991) 194; Fl. Intramong. ed. 2, 4 (1993) 644; Opred. rast. Sr. Azii [Key to Plants of Mid. Asia] 10 (1993) 561.

Described from East. Kazakhstan (Balkhash Region). Type in St.-Petersburg (LE).

On hummocky sand, exposed sandstones, in depressions between dunes, saxaul thickets.

IA. Mongolia: *Alash. Gobi* (Alashan, Urat – Fl. Intramong. (1993) l.c.).

IIA. Junggar: *Dzhark.* (right bank of Ili river, 7–8 km south-west of Suidun settlement on road to Santokhodze pier, hummocky-ridgy sand, Aug. 31, 1957 – Yun., Li et Yuan').

General distribution: Aralo-Casp., Fore Balkh., Jung.-Tarb.; endemic in Cent. Asia.

107. A. sphaerocephala Krasch. in Acta Inst. Bot. Acad. Sci. URSS, 1 (1936) 348; Grub. Konsp. fl. MNR [Conspectus of Flora of Mongolian People's Republic] (1955) 269; Leonova in Grub. Opred. rast. Mong. [Key to Plants of Mongolia] (1982) 251; Fl. Intramong. 6 (1982) 121; Fl. Sin. 76, 2 (1991) 189; Fl. Intramong. ed. 2, 4 (1993) 641; Gub. Konsp. fl. Vneshn. Mong. [Conspectus of Flora of Outer Mongolia] (1996) 99. – Ic.: Fl. Intramong. (1982) tab. 42, fig. 1–7; Fl. Intramong. (1993) tab. 251, fig. 7–12.

Described from Mongolia (Alash. Gobi). Type in St.-Petersburg (LE). Map 7.

On thin, saline, rubbly and ridgy sand, solonchak lowlands and depressions between dunes, on pebble beds, rocky and rubbly slopes of low mountains.

IA. Mongolia: *Depr. Lakes* (Gub. l.c.), *Val. Lakes* (Bayan-Gobi, Shar-Khubsugul area, rocky slopes of conical hillocks, Aug. 9, 1943 – Tsegmid; Dzhinsetu somon, Burungul-Khongor area, steppe, Aug. 17, 1943 – Yun.; Bayan-Tsagan somon road to Delger somon, desert steppe, Aug. 28, 1948 – Grub.; 40 km north-east of Buyan-Tsagan somon to Bayan-Khongor, consolidated sand, Aug. 4, 1983 – Gub.), *Gobi Alt.* (between Dzolin and

Bayan-Tsagan mountains, south of Bayan-Tukhum lake, Sept. 5, 1931 — Ik.-Gal., syntypus!; south. trail of Artsa-Bogdo-nuru on road to Leg somon, sandstones with saxaul, July 13; same site, near Khatsar-Khuduk collective, sandstones, July 23–1948, Grub.; 5 km east-north-east of Noyan somon, outcrops of conglomerates, July 25, 1972 — Guricheva et Rachk.; Tostu mountain range, Aug. 24, 1982 — Gub.), *East. Gobi* (Ail'-Bayan somon, 12 km south-east of Ulegei-khilda, sand, July 2, 1941 — Yun.; 115 km south-east of Sain-Shanda, ridgy sand, July 21, 1971 — Isach. et Rachk.; 8 km south of Nomogon somon, sand, June 20; 50 km north-east of Bulgan somon, sandy desert, Aug. 1–1972, Rachk. et Guricheva; 90 km south — south-east of Dalan-Dzadagad town, sand-covered foothills, Aug. 31, 1982 — Gub.), *West. Gobi* (20–30 km south-east of Atas-Bogdo mountain range and 15 km south of Khukhu-usu area, thin sand, Aug. 11; Tsagan-Bogdo mountain range, nor. slope, Aug. 3–1943, Yun.; 20 km south-east of Suchzhi-Khuduk well, plain, July 17, 1972 — Rachk. et Guricheva), *Alash. Gobi* (Noyan somon, 15–20 km west of Obotu, shrubby steppe along gorge, July 28, 1943 — Yun.; near Bumbutui well, Alashan, Sept. 4, 1880 — Przew., typus!; Edzin-gol river, between Khair-Toor area and Tsagan-Buryuk, July 18; Tszazh-Shar-khulusun area, Aug. 10–1886, Pot., syntypus!; 37 km north-east of Bayan-Khoto town, depressions between dunes, June 30, 1958 — Petr.; 110 km south — south-east of Nomogon somon, hummocky consolidated sand, July 3, 1981; Severei somon, Khon-Goryn-Els sand, Aug. 27, 1982 — Gub.; Alashan — Fl. Intramong. (1993) l.c.), *Khesi* (8–9 km west of Gaotai, thin sand, Oct. 8, 1957; from Uvei to Manchantszin', in depressions, June 23; 18 km west of Rzhan'e town, Kheikhe river high terrace, July 13; 50 km north-west of Tszin't town, south-east. extremity of desert, July 19; 60 km north-east of Tszin'ta town, arid bed, July 23; 36 km north of An'si settlement, pebbly-sandy plain, July 27; 17 km north of An'si settlement, arid bed, July 27; Kumtag, on highway to Qaidam, depressions between dunes, Aug. 4–1958, Petr.), *Ordos* ("Urat, Khodtsy, Denkou" — Fl. Intramong. (1993) l.c.).

IIA. Junggar: *Jung. Gobi* (20 km north-east of Bodonchiin-Khure, wormwood-sand desert, Aug. 18, 1947 — Yun.; 14 km from Bodonchiin-gol on Guchen highway, on sand, Sept. 19; 17 km west — north-west of head of Ubguchuin-gol, in hummocky area, Sept. 9– 1948, Grub.; 55 km north of Gan'khetsze, sand dunes, No. 1920, Sept. 27, 1957 — Kuan; 30 km north-east of Burchum to Shara-sume, thin sand, July 5; 5–8 km north of Bului-Tokhei settlement, in lower courses of Urumchi river, ridgy sand, July 9–1959, Yun. et Yuan').

General distribution: China (Nor., Nor.-West.), Nor. Mong. (Hang.).

108. **A. subdigitata** Mattf. in Feddes Repert. 22 (1926) 243; Kitag. Lin. Fl. Mansh. (1939) 433; Grub. in Bot. zhurn. 12 (1976) 1953; Leonova in Grub. Opred. rast. Mong. [Key to Plants of Mongolia] (1982) 245; Fl. Intramong. 6 (1982) 113; Pl. vasc. Helansch. (1986) 357; Gub. Konsp. fl. Vneshn. Mong. [Conspectus of Flora of Outer Mongolia] (1996) 99. — *A. subdigitata* var. *thomsonii* (Pamp.) S.J. Hu in Fl. Xizang. 4 (1985) 783. — *A. dubia* Wall. ex Bess. var. *subdigitata* (Mattf.) Y.R. Ling in Kew Bull. 42, 2 (1989) 445; Fl. Intramong. ed. 2, 4 (1993) 662. — Ic.: Fl. Intramong. (1982) tab. 40, fig. 6, 7 and (1993) tab. 257, fig. 6, 7.

Described from China. Type in St.-Petersburg (LE). Map 7.

In steppes, on steppe rubbly slopes of mounds and conical hillocks, wormwood-cereal grass and cereal grass-forbs steppe associations, on meadows, along gorge floors, in juniper forests.

IA. Mongolia: *Mong. Alt.* (pass through Dobpig Khuren-Nuru into Mong. Altay, July 11, 1947 — Yun.), *Depr. Lakes* (Adzhi-Bogdo mountain range, midportion of Mannaitu-ul creek valley, Aug. 7, 1947 — Yun.), *East. Mong.* (Khailar town, in sandy site, July 20; same site, Sishan' mounds, in forest, July 29, 1951 — Wang), *Gobi Alt.* (east. extremity of Dundu-Saikhan mountain range, July 22; 1–2 km south of Noyan somon, slopes of low conical hillocks, Aug. 25; nor. slope of Ikhe-Bogdo mountain range, juniper forests along steep slopes, Sept. 12; same site, upper course of Nokhoitin-Khundei creek valley, mountain steppe, Sept. 14–1943, Yun.; west. portion of Nemegetu-Nuru mountain range, floor of large ravine 2 km from estuary, Aug. 4; Tostu-Nuru mountain range, south. slope of gorge of main ravine, sandy-pebbly floor, Aug. 15; same site, on nor. slope 5 km from estuary, Sept. 5–1948, Grub.; 8 km south-east of Tost somon, near Khugshu-ul, in gorge, June 28, 1974 — Rachk. et Volk.), *West. Gobi* (100 km west — south-west of Ekhiin-gol oasis, near spring, Aug. 26, 1976 — Rachk. et Damba), *East. Gobi* (Ikh-Shantai-Nuru mountains, in crevices, Aug. 30, 1976 — Guricheva et Rachk.; near Shara-Khulusun-bulak spring, Aug. 4, 1978 — Lobachev), *Alash. Gobi* (Gub. l.c.).

IIA. Junggar: *Jung Gobi* (Baitag-Bogdo mountain range, Ulyasutai gorge, 3 km from estuary, Sept. 18, 1948 — Grub.).

IIIA. Qinghai: *Nanshan* (Tetung mountain range, July 26, 1880 — Przew.; on way to Kuku-nor lake from Alashan, Aug. 5, 1909 — Czet.; 33 km west of Xining town, rocky wormwood-cereal grass steppe, Aug. 5; 66 km west of Xining town, cereal grass-forbs mountain steppe, Aug. 5 — 1959, Petr.).

General distributioon: Far East, Nor. Mong. (Hang., Mong.-Daur.), China (Nor.-East.).

109. A. tomentella Trautv. in Bull. Soc. natur. Moscou, 39, 1 (1896) 351; Krasch. in Kryl. Fl. Zap. Sib. 11 (1949) 2772; Poljak. in Fl. SSSR, 26 (1961) 554; Fl. Kazakhst. 9 (1966) 108; Opred. rast. Sr. Azii [Key to Plants of Mid. Asia] 10 (1993) 562; Gub. Konsp. fl. Vneshn. Mong. [Conspectus of Flora of Outer Mongolia] (1996) 99.

Described from Cent. Kazakhstan (Chu river basin and Sarysu). Type in St.-Petersburg (LE).

In arid steppes and deserts on light loamy sand soils, sand, loam, rarely in chee grass thickets.

IA. Mongolia: *Khobd.* (Gub. l.c.); *Depr. Lakes* (Margats mountains, 4–5 km north-east of Santa-Mvergats somon, sandy steppe with wormwood, Aug. 19, 1944; Borig-Del' sand south-east of Bayan-nur lake, July 25, 1945 — Yun.; same site, 44 km north-east of state farm settlement on road to Tes somon, steppe, July 21, 1974 — Banzragch et Munkhbayar; same site, 50 km north-west of Tes somon, steppe, June 12, 1978 — Karam. et Beket.).

IIA. Junggar: *Jung. Gobi* (lower course of Uinchi river, 15-20 km before wintering site of somon, vicinity of solonchak-like meadow, Aug. 29, 1947 — Yun.; left bank of Chernyi Irtysh river, 20 km beyond Shipoti settlement on road to Kokh-Togai from Burchum, desert, July 8; Urunchu river valley, 4–5 km south of Bulun-Tokhoi settlement, hummocky chee grass thicket, July 9, 1959 — Yun. et Yuan').

General distribution: Aralo-Cast, Fore Balkh., Jung.-Tarb., West. Sib., Nor. Mong. (Fore Hubs., Hang.).

110. A. waltonii Drumm. ex Pamp in Nouv. Giorn. Bot. Ital. n.s. 34 (1927) 707; Enum. vasc. pl. Xizang. (1980) 365; Fl. Xizang. 4 (1985) 791; Fl. Sin. 76, 2 (1991) 205.

116

Described from South. Tibet. Type in Firenze (Florence) (FI). Isotype in St.-Petersburg (LE).

In mountain steppes.

IIIB. Tibet: *Weitzan* (Burkhan-Budda mountain range, Nomokhun-gol river, July 3–15, 1884 – Przew.), *South.* (Gyangtze, July–Sept. 1904 – Walton, isotypus!; "Lhasa, Nan'mulin, Tszyantszy, Lankha-tsza" – Fl. Xizang. (1985) l.c.).

General distribution: Himalayas (Kashmir).

111. A. wellbyi Hemsl. et Pears. in Hemsl. Fl. Tibet (1902) 183; Enum. vasc. pl. Xizang. (1980) 365; Fl. Xizang. 4 (1985) 789; Fl. Sin. 76, 2 (1991) 204. – *A. salsoloides* auct. non Willd.: Pamp. Fl. Carac. (1930) 209.

Described from Tibet. Type in London (K).

In alpine and subalpine mountain belts.

IIIB. Tibet: *Chang Tang* ("sandy gravelly soils in valleys at 17,100 ft. 1891 – Thorold; 86°10'; 35°19': 16,214 ft. July 16, 1896 – Wellby et Malcolm, typus!; near Mangtsa Tso, 82°08', 34°48', 17,800 ft. July 27, 1896 – Deasy et Pike" – Hemsley, l.c.; "Shuan-khu, Gaitsze" – Fl. Xizang. 1985, l.c.), *South.* ("Lhasa, Shiganze, Tszyatszy, Chzhunba, Pulan'" – Fl. Xizang. 1985), l.c.).

General distribution: endemic in Cent. Asia.

112. A. wudanica Liou et W. Wang in Acta Phytotax. Sin. 17 (1979) 188; Fl. Intramong. 6 (1982) 120; Fl. Sin. 76, 2 (1991) 191; Fl. Intramong. ed. 2, 4 (1993) 641; Gub. Konsp. fl. Vneshn. Mong. [Conspectus of Flora of Outer Mongolia] (1996) 100. – Ic.: Fl. Intramong. (1982) tab. 42, fig. 1–6; Fl. Intramong (1993) tab. 250, fig. 1–6.

Described from Inner Mongolia. Type in Beijing (PE).

On semiconsolidated sand, in ravines.

IA. Mongolia: *Cent. Khalkha* (Gub. l.c.), *East. Mong.* (10 km east of Dariganga somon, Ganza-nur lake, sand, Aug. 25, 1980 – Dar.), *East. Gobi* (Khotan-Bulak somon, 18 km north-west of Khetsu-ul mountains, Aug. 25; Bayan-Obo somon, 10 km north-west of Gashun-Sukhaiga post, Aug. 27–1980, Dar.), *Alash. Gobi* (Bayannor ajmaq, Urat – Fl. Intramong. (1993) l.c.).

General distribution: China (East.).

113. A. xanthochroa Krasch. in Not. syst. (Leningrad), 9 (1946) 174; Grub. Konsp. fl. MNR [Conspectus of Flora of Mongolian People's Republic] (1955) 269; Leonova in Grub. Opred. rast. Mong. [Key to Plants of Mongolia] (1982) 251; Fl. Sin. 76, 2 (1991) 199; Fl. Intramong. ed. 2, 4 (1993) 648. – *A. arenaria* auct. non DC.: Fl. Intramong. 6 (1982) 123.

Described from Mongolia (East. Gobi). Type in St.-Petersburg (LE).

In arid steppes and deserts, saxaul forests, on hummocky and dune, rarely solonetzic sand, pebble beds in ravines, rubbly-sandy trails and rocky slopes of mountains.

IA. Mongolia: *Khobd.* (east. slope of Dzhindzhilin pass, steppe, July 30, 1945 – Yun.), *Mong. Alt.* (pass on road from Bayan-Tsagan somon to Dzakhoi, through east. slope of

Mong. Altay, desert steppe, Aug. 20, 1943 – Yun.), *Cent. Khalkha* (from Choiren to Naran area, Sept. 5, 1927 – Zam.; Choiren-ula, 1940 – Sanzha; old Ulan-Bator – Dalan-Dzadagad road, 15 km northward, steppe with pea shrub, July 13, 1948 – Grub.; 170 km from Undurkhan on highway to Erentsab, Ul'tsza river valley, steppe, July 27, 1949 – Yun.), *East. Mong.* (vicinity of Baishintin-sume, Aug. 18, 1927 – Zam.; same site, Sept. 9, 1930 – Pob.; Dariganga, 25 km north-east of Argaleul mountain, Sept. 1; 40 km north-east of Argaleul mountain, near Ovansokul' well, arid bed, Sept. 5; same site, 25 km west of Baishintu-sume, near Bagamotne-bulak spring, Sept. 9–1931, Pob.; Moltsok-Elisu sand, 3–4 km north of Moltsok-Khida, semiconsolidated sand, June 16, 1944; 45 km from Shara-Bulak lake, on road to Tamtsak-bulak, Aug. 4, 1949 – Yun.; Argalantu mountains, Tsagan-nor knoll, near Bayan-Munku-khida ruins, ravine floor, Aug. 6, 1970 – Grub. et al), *Depr. Lakes* (Shargin-Gobi, pass from Shargin-Tsagan-nur lake to Khasagtu-Khairkhan mountain range, trail, Sept. 14, 1930 – Pob.), *Val. Lakes* (Tuin-gol river, steppe along ridges, Sept. 1, 1924 – Pavl.; 45 km south of Tugreg settlement, vicinity of Khongoryn-Tala sand, July 28, 1983 – Gub.), *East. Gobi* (6 km east – north-east of Del'girkh somon, solonchak derris area, Sept. 11; 72 km north-east of Sain-Shanda town, sand, Sept. 13– 1940, Yun.; 40 km north of Altyn-Shire, granitic hummocky area, Aug. 6, 1971 – Isach. et Rachk.; 75 km north-west of Sain-Shanda town, hummocky area, July 31, 1974 – Rachk. et Volk.; 25 km east of Del'girkh settlement, on residual hillocks, July 30, 1981 – Gub.; nor.-east of Bulgan somon, Bayan-deag aeram, Aug. 11; vicinity of Mandal-obo, Aug. 11– 1984, Kam. et Dar.; "Ulanpab, Urot" – Fl. Intramong. (1993) l.c.), *Gobi Alt.* (Gub. l.c.), *Alash. Gobi* ("Alashan" – Fl. Intramong. (1993) l.c.).

IIA. Junggar: *Jung. Gobi* (Ulyungur lake, Aug. 3, 1876 – Pot.; Bodonchiin-gol on road to Bodonchiin-Baishing from Tamcha somon, arid bed, Sept. 8–10, 1948 – Grub.; plain 2–3 km west of Arasan settlement on road to Tamirtam, sand-covered terrace, Aug. 6, 1957 – Yun. et al).

General distribution: Nor. Mong. (Hang., Mong.-Daur., Fore Hing.).

114. A. xigazeensis Ling et Y.R. Ling in Acta Phytotax. Sin. 4, 18 (1980) 511; Enum. vasc. pl. Xizang. (1980) 365; Fl. Xizang. 4 (1985) 792. – *A. salsoloides* auct. non Willd.: Hemsl. Fl. Tibet (1902) 183; p.p.; Pamp. Fl. Carac. (1930) 209, p.p., excl. syn. *A. wellbyi* Hemsl. et Pears. and var. *prattii* Pamp. – Ic.: Fl. Xizang. (1985) tab. 347, fig. 1–9.

Described from Tibet. Type in Beijing (PE).

In high mountains.

IC. Qaidam (Tsakei-usse Tal sudostlich von Tsaidam, 3700 m alt, June 30, 1906 – Tafel).

IIIB. Tibet: *Chang Tang* ("Shuankhu" – Fl. Xizang. (1985) l.c.); *South.* ("Lhasa" – Fl. Xizang. (1985) l.c.).

General distribution: China (Tibet).

118   115. A. xylorhiza Krasch. ex Filat. in Bot. zhurn. 71, 11 (1986) 1553; Krasch. in Mat. po ist. fl. i rast. SSSR, 2 (1946) 155, nom. nud.; Grub. Konsp. fl. MNR [Conspectus of Flora of Mongolian People's Republic] (1955) 270; Gub. Konsp. fl. Vneshn. Mong. [Conspectus of Flora of Outer Mongolia] (1996) 100.

Described from Mongolia (Cent. Khalkha). Type in St.-Petersburg (LE). Plate V, fig. 4. Map 7.

On rubbly, hummocky and semiconsolidated sand, sandstone outcrops along gorge floors.

IA. Mongolia: *Mong. Alt.* (slope in Indertiin-gol river valley, steppe, Aug. 25, 1947 — Yun.), *Cent. Khalkha* (Dzhargalante river basin), waterdivide between Ara-Dzhargalante and Ubur-Dzhargalante, rubbly sand, Aug. 10, 1925 — Krasch. et Zam., typus!; same site, sand around Uste mountain, Aug. 17, 1925 — Krasch. et Zam., paratypus!; 12 km north-west of Delger-Hangay mountains, sand mounds, Aug. 6, 1970 — Banzragch, Karam, Lavr.), *East. Mong.* (Moltsok-Elis sand, 3–4 km north of Moltsok-Khida, hummocky, semiconsolidated sand, June 16, 1944 — Yun.), *Depr. Lakes* (Gub. l.c.), *Val. Lakes* (east. extremity of Chuilin-Tal plain, sandy loam, Aug. 26, 1943 — Yun.), *Gobi Alt.* (20 km north-west of Shine-Zhinst somon, June 27, 1974 — Rachk.), *East. Gobi* (60 km east of Ulan-Baderkhu somon centre, semiconsolidated sand, June 19; 16 km north-east of Sain-Shanda town, in derris scrubs, Aug. 1; 30 km south-west of Sain-Shanda, steppified saltwort desert, Sept. 18–1940, Yun.; Erdene somon, south. extremity of Khalzan-Bilyuta mountain range, desert steppe, June 15, 1941 — Yun.; 9 km south-west of Ikhe-Dzhargalanta, July 15; 20 km south-west of Saikhan-Dulun, along gorge floor, July 18; 30 km south-east of Hubsugul, on sandstone outcrops, July 27; 40 km north — north-east of Sain-Shanda, Aug. 3–1971, Isach. et Rachk.; 30 km north-west of Bulgan, desert, Aug. 7, 1972 — Isach.), *Alash. Gobi* (nor. extremity of Gashuin-nur lake, July 27, 1943 — Yun.).

General distribution: Nor. Mong. (Hang., Mong.-Daur); endemic.

Subgenus 3. Seriphidium (Bess.) Peterm.

Deutsch. Fl. (1848) 294. — *Artemisia* sect. *Seriphidium* Bess. in Bull. Soc. natur. Moscou, 1 (1828) 222. — *Seriphidium* (Bess.) Poljak. in Trudy Inst. Bot. Ac. Nauk. Kazakhsk. SSR, 11 (1961) 171

1. Leaves tri- or palmatisected (excepting uppermost); subshrubs pubescent with short straight appressed hairs ....
.... 127. A. juncea Kar. et Kir.

+ Leaves of lower cauline and barren shoots pinnatilobed or pinnatisected; perennials and subshrubs, pubescent with flexuose cobwebby braided hairs or glabrous .... 2.

2. Lower and middle cauline leaves pinnatilobed; leaf blade elongated, lateral lobes contracted, poorly developed, entire, dentate or incised into 2–3 small orbicular lobules ....
.... 132. A. santolina Schrenk.

+ Lower and middle cauline leaves pinnatisected; leaf blade ovoid or orbicular, its lateral lobes well-developed, pinnatisected .... 3.

3. Herbaceous perennials with few shoots, somewhat lignifying at base .... 4.

+ Small subshrubs with many contracted shoots highly lignifying at base .... 9.

4. Fertile shoots rod-shaped, branched from base, with erect or somewhat curved branchlets; anthodia oblong-ovoid, cuneately narrowed toward base, form loose or subglobose panicle .... 5.

+ Fertile shoots stout, straight, branched from centre, with straight, obliquely standing branchlets; anthodia ovoid, not narrowed toward base, form pyramidal compact panicle .... 7.

5. Leaf blade dark green above, foveolate-punctate-glandular with 4-6 pairs of proximated lobules; terminal leaf lobules with chondroid cusp at tip; anthodia up to 2 mm long, single or in groups of 2-3 on lateral, up to 7 cm long, branchlets of panicle .... 116. A. assurgens Filat.

+ Leaf blade greyish on both sides, without glandular pubescence, with 2-4 pairs of highly developed lobules; terminal leaf lobules obtuse at tip or short-cuspidate; anthodia 3-5 mm long, single, on lateral, up to 15 cm long, branchlets of panicles .... 6.

6. Lower cauline leaves bipinnatisected; terminal leaf lobules somewhat cuspidate at tip; anthodia sessile, form narrow or broad pyramidal panicle; phyllary persistent on ripening of achene .... 122. A. gobica (Krasch.). Grub.

+ Lower cauline leaves 1-2-pinnatisected; terminal leaf lobules obtuse at tip; anthodia short-stalked, form subglobose panicle; phyllary deciduous on ripening of achenes ....
.... 128. A. kaschgarica Krasch.

7. Plant with compact white-tomentose pubescence until end of vegetation; lower cauline leaves with long petioles surpassing leaf blade; terminal leaf lobules spatulately broadened at tip .....
.... 134. A. schrenkiana Ledeb.

+ Plant greyish green or brown toward end of vegetation; lower cauline leaves on petioles as long as leaf blade or shorter; terminal leaf lobules short-cuspidate at tip .... 8.

8. Plant with slender creeping rhizome; lower cauline leaves bipinnatisected; terminal leaf lobules linear; plant of steppe zone .... 130. A. nitrosa Web. ex Stechm.

+ Plant with stout, vertical, woody root; lower cauline leaves 1-2 pinnatisected; terminal leaf lobules broad-lanceolate. Desert plant .... 131. A. saissanica (Krasch.) Filat.

9. Inflorescence spicate-paniculate; anthodia in small numbers aggregated into compact glomerules or compact, oval, spaced or proximate, spikelets .... 10.

+   Inflorescence paniculate; anthodia proximated or spaced on lateral branchlets of panicle, not forming compact glomerules or spikelets .... 11.

10.   Blades of lower cauline leaves orbicular-ovoid in profile; terminal leaf lobules linear-subulate, 5–7 (9) mm long; 5–7 (9) anthodia gathered into compact glomerules forming spicate inflorescence .... 135. A. skorniakovii Winkl.

+   Blades of lower cauline leaves oblong-ovoid in profile; terminal leaf lobules linear or narrow-lanceolate, up to 5 mm long; 3–5 anthodia gathered into compact spikelets forming interrupted-spicate inflorescence .... 117. A. borotalensis Poljak.

11.   Lower cauline leaves and leaves of barren shoots invariably 1–2-pinnatisected, i.e. lower lateral lobules entire and upper ones once again laciniated .... 12.

+   Lower cauline leaves and leaves of barren shoots 2–3-pinnatisected .... 13.

12.   Plant up to 10 (15) cm tall; inflorescence. a racemose panicle with short, somewhat declinate lateral branchlets ....
.... 133. A. schischkinii Krasch.

+   Plant 15–30 cm tall; inflorescence, a loose, diffuse panicle with subhorizontally declinate lateral branchlets ....
.... 138. A. terrae-albae Krasch.

13.   Lower cauline leaves and leaves of barren shoots 2–3-pinnatisected .... 14.

+   Lower cauline leaves and leaves of barren shoots only bipinnatisected .... 16.

14.   Fertile stalks many, slender, up to 30 cm tall; leaves of barren shoots and lower cauline leaves with petioles greatly enlarged toward base; phyllary with fine cobwebby pubescence and punctate-glandular .... 124. A. gracilescens Krasch.

+   Fertile shoots not many, 30–40 (45) cm tall; leaves of barren shoots and lower cauline leaves on petioles not enlarged toward base; phyllary with cobwebby tomentose pubescence ....
.... 15.

15.   Fertile shoots compactly leafy; leaves generally persist until end of vegetation; panicle with short lateral, up to 5 cm long branchlets; anthodia form compact spikelets ....
.... 119. A. elongata Filat. et Ladyg.

+   Fertile shoots somewhat leafy; panicle with 5–10 cm long lateral branchlets; anthodia form loose spikelets ....
.... 118. A. compacta Fisch. ex DC.

16. At beginning of vegetation, plants greyish, with fine cobwebby pubescence; greenish, yellowish brown, brown toward end of vegetation .... 17.

\+ Until end of vegetation, plants whitish or greyish due to compact tomentose or cobwebby-tomentose pubescence ...... 20.

17. Fertile shoots greenish or yellowish brown toward end of vegetation, branched almost up to centre; lower cauline leaves stalked, as long as leaf blade or somewhat shorter .... 18.

\+ Fertile shoots brown or dark brown toward end of vegetation, branched in upper part; lower cauline leaves petiolate, considerably surpassing blade .... 19.

18. Fertile shoots somewhat stout, 45–70 cm tall, ascending at base; panicle broad-pyramidal, with up to 15 cm long lateral branchlets; anthodia up to 2 mm long, oblong-ovoid; phyllary with somewhat cobwebby pubescence, inner phyllary with broad golden border .... 121. A. fulvella Filat. et Ladyg.

\+ Fertile shoots slender, up to 30 (35) cm tall, not ascending at base; panicle narrow-pyramidal, with up to 6 (8) cm long lateral branchlets; anthodia 2–3 mm long, ovoid; phyllary subglabrous, inner phyllary with narrow brownish border ....
.... 120. A. fedtschenkoana Krasch.

121 19. Fertile shoots 25–30 (35) cm tall; leaf blade profile ovoid; terminal leaf lobules narrow-linear, 3–5 mm long; anthodia gathered 3–5 on short, up to 3 (5) cm long lateral branchlets of panicle .... 123. A. gorjaevii Poljak.

\+ Fertile shoots 30–45 cm tall; leaf blade profile oblong-oval; terminal leaf lobules 5–12 mm long, filiform-linear; anthodia single or gathered 2–3 on much longer (5–10 cm long) lateral branchlets of panicles ....
.... 137. A. sublessingiana Krasch. ex Poljak.

20. Fertile shoots grow from highly lignified base (to 10–15 cm height); phyllary of anthodia with broad, golden border .... 21.

\+ Fertile shoots grow from base, lignified to a shorter height (not more than 10 cm); phyllary of anthodia with narrow scarious brownish border .... 22.

21. Plant up to 30 cm tall, greyish until end of vegetation due to compact cobwebby-tomentose pubescence; leaf blade ovoid or suborbicular in profile; terminal leaf lobules linear or linear-lanceolate, subobtuse at tip; panicle narrow-pyramidal, compact, with short lateral branchlets, with 3–5 anthodia gathered on them in compact groups ....
.... 136. A. subchrysolepis Filat.

+ Plant 30–40 cm tall, with flocculose-tomentose pubescence toward end of vegetation; leaf blade oblong-ovoid in profile, terminal leaf lobules lanceolate, short-cuspidate at tip; panicle broad-pyramidal, loose, with much longer lateral branchlets; anthodia single or 2–3 of them together ....
.... 129. A. mongolorum Krasch.

22. Fertile shoots branched from base; lower cauline leaves on petioles surpassing blade; terminal leaf lobules lanceolate; panicle loose, anthodia ovoid, sessile ....
.... 126. A. issykkulensis Poljak.

+ Fertile shoots branched in upper portion; lower cauline leaves on petioles as long as blade; terminal leaf lobules linear; panicle compact; anthodia oblong-ovoid, short-stalked ....
.... 125. A. heptapotamica Poljak.

116. A. assurgens Filat. in Novit. syst. pl. vasc. 19 (1982) 178; id. 23 (1986) 236; Gub. Konsp. fl. Vneshn. Mong. [Conspectus of Flora of Outer Mongolia] (1996) 96. — Ic.: Filat. (1986) Plate 9, fig. 3, 4.

Described from Mongolia (West. Gobi). Type in St.-Petersburg (LE). Plate VI, fig. 5. Map 8.

In solonchaks, saline rocky-clayey soils, solonchak meadows, gorge floors, vicinity of springs, rarely on saline sand.

IA. Mongolia: *Mong. Alt.* (Uenchi river basin, valley of Nariin-gol river, 9 km south of Ikhe-Ulan-daba pass, Aug. 13, 1979—Grub., Muld. et Dar.; 35 km east of Beger settlement, Aug. 8, 1983—Gub.), *Gobi Alt.* (nor. slope of Tostu-ul mountain range, Naran-Bulak, July 16, 1973—Golubk. et Tsogt; intermontane plain south of Altan-ul (Nemegetu mountain range), near Naran-Dats spring, Oct. 5, 1979—Grub., Muld. et Dar.), *West. Gobi* (Dzakhoi-Dzaram area, south. Tsel' somon, Aug. 15, 1943—Yun.; Khatyn-Suda collective, Aug. 6, 1946—Yun.; 50 km south of Ekhiin oasis on road to Segs-Tsagan-Bogdo town, Aug. 18; 100 km west of Ekhiin-gol oasis, Aug. 27; 15 km east of Altay somon, Aug. 27—1973, Isach. et Rachk., typus!; south-west of Ekhiin-gol, Tsagan-Burchasyn-bulak spring, Aug. 16, 1976—Rachk.), *Alash. Gobi* (15 km north-west of Dzin'ta town, on old bed of Beitakhe river, July 18, 1958—Petr.).

IIA. Junggar: *Jung. Gobi* (Bodonchiin-gol floodplain, lower courses, 15 km south-west of Altay settlement, Aug. 1, 1979—Gub.).

General distribution: endemic in Cent. Asia.

Note. Species close to *A. mongolorum* Krasch. differing from it in slender, highly procumbent shoots, faint, fine, cobwebby-tomentose pubescence punctate-glandular pubescence of leaves and small anthodia. Typical halophyte.

117. A. borotalensis Poljak. in Not. syst. (Leningrad), 16 (1954) 425; Filat. in Novit. syst. pl. vasc. 21 (1984) 178; ibid, 23 (1986) 236; Opred. rast. Sr. Azii [Key to Plants of Mid. Asia] 10 (1993) 565; Gub. Konsp. fl. Vneshn. Mong. [Conspectus of Flora of Outer Mongolia] (1996) 97. — *Seriphidium borotalense* (Poljak.) Ling et Y.R. Ling, Fl. Sin. 76, 2 (1991) 281. — Ic.: Filat. (1986) Plate 10, fig. 5, 6.

Described from Jung. Ala Tau (Borotala river). Type in St.-Petersburg (LE).

On rocky-clayey and rubbly-clayey slopes of mountains, in valleys of mountain rivers and lake basins. Dominant species of alpine wormwood-cereal grass steppes, forms plant associations at 2200–3200 m above sea.

IA. Mongolia: *Mong. Alt.* (Khalyun area, upper mountain belt, Aug. 24; Bus-Khairkhan mountain range, July 17–1943, Yun; Shad-zgain-nuru mountain range, valley of Ulyasutai-gol river, June 27, 1973–Golubk. et Tsogt; south. slope of Mong. Altay, July 31, 1977–Rachk.; valley of Nariin-gol river, 9 km south of Ikhe-Ulan-daba pass, Aug. 13, 1979–Grub., Muld. et Dar.; valley of Sangiin-gol river, upper course of Dzhelta river, Aug. 13, 1979–Gub.; Bulgan river basin, midcourse of Bayan-gol river, July 24, 1984–Kam. et Dar.).

IIA. Junggar: *Cis-Alt.* (Qinhe, Aug. 1, 1956–Ching), *Jung. Alat.* (Borotala river, Aug. 6, 1978–A. Reg., typus!; Maili mountain range, south-west. vicinity toward Karaganda pass, Aug. 14; from Borotal to Sairam-Nur lake, Aug. 18; basin of Kuitun river, between Borotal and Sairam-Nur lake, Aug. 18; same site, Bain-gol creek valley, Aug. 29–1957, Yun. et al), *Tien Shan* (Sairam lake, July 1877–A. Reg.; 3 km south of Nyutsyuan'tsza, July 17; nor. bank of Barkul' lake, Sept. 6–1957, Kuan; 6 km south-west of Sayatai-nur settlement, Aug. 19; basin of Sairam lake on road to Talki (Sairam) pass, Aug. 19; upper course of Tekes river, Aug. 24–1957, Yun.; valley of Muzart river in its upper course, Sept. 17; Sogdan-Tau mountain range, north-west of Kokshal-Dar'ya river valley, Sept. 19 –1958, Yun., Li et Yuan'), *Jung. Gobi* (Baitag-Bogdo mountain range, Aug. 8; Ikh-Khavtag mountain range, Aug. 12, 1977–Isach. et Rachk.).

General distribution: Jung.-Tarb., Nor. Tien Shan, endemic in Cent. Asia.

118. A. compacta Fisch. ex DC. 6 (1838) 102; Krasch. in Kryl. Fl. Zap. Sib. 11 (1949) 2784; Grub. Konsp. Fl. MNR [Conspectus of Flora of Mongolian People's Republic] (1955) 263; Poljak. in Fl. SSSR, 26 (1961) 585, excl. syn. *A. tianschanica* Krasch. ex Poljak.; Fl. Kirgiz. 11 (1965) 183; Leonova in Grub. Opred. rast. Mong. [Key to Plants of Mongolia] (1982) 253; Fl. Intramong. 8 (1985) 322; Filat. in Novit. syst. pl. vasc. 23 (1986) 236; Fl. Intramong. ed. 2, 4 (1993) 666; Opred. rast. Sr. Azii [Key to Plants of Mid. Asia] 10 (1993) 565; Gub. Konsp. fl. Vneshn. Mong. [Conspectus of Flora of Outer Mongolia] (1996) 96. — *A. albida* auct. non Fisch. ex DC.: Fl. Kazakhst. 9 (1966) 139. — *Seriphidium compactum* (Fisch. ex DC.) Poljak. in Trudy Inst. Bot. Ac. Nauk Kazakhsk. SSR, 11 (1961) 175; Fl. Sin. 76, 2 (1991) 284. — Ic.: Filat. (1986) Plate 8, Fig. 15, 16.

Described from Altay. Type in St.-Petersburg (LE).

Solonetzic desert steppes, saline loam, rarely on rocky and rubbly slopes of low mountains and knolls. East. Kazakhstan species, dominant in solonetzic desert steppes.

123   IA. Mongolia: *Khobd.* (nor. bank of Tsagan-nur lake, Aug. 8–9, 1972–Metel'tseva), *Mong. Alt.* (Saksany-gol river, July 31, 1909–Sap., Khara-Adzarga mountain range, valley of Shutyn-gol river, Aug. 28, 1930–Pob.; nor. slope of Mong. Altay, lower mountain belt, Aug. 24, 1943–Yun.), *Cent. Khalkha, Depr. Lakes, Val. Lakes* (Gub. l.c.).

IIA. Junggar: *Cis-Alt.* (15 km nor.-west of Shara-Sume, July 7, 1949–Yun. et al.), *Jung. Gobi* (Gub. l.c.).

142

General distribution: Fore Balkh. (nor.), Jung.-Tarb., West. Sib. (Altay), Nor. Mong. (Hang.).

119. A. elongata Filat. et Ladyg. in Novit. syst. pl. vasc. 18 (1981) 225; Filat. in Novit. syst. pl. vasc. 21 (1984) 178; ibid 23 (1986) 236; Opred. rast. Sr. Azii [Key to Plants of Mid. Asia] 10 (1996) 566. — Ic.: Filat. (1986) Plate 11, fig. 7, 8.

Described from Tien Shan (Tersk Ala Tau). Type in St.-Petersburg (LE).

On rubbly-rocky, rubbly-clayey and clayey-loessial slopes of mountains, in intermontane valleys. Dominant species of midmountain wormwood-cereal grass with pea shrub steppes, 1600–2600 m above sea.

IB. Kashgar: *Nor.* (Uch-Turfan, Kukurtuk gorge, June 27, 1908 – Divn.), *West.* (53 km west – north-west of Kashgar, on road to Ulugchat, July 17; 4-5 km west – north-west of Kzyl-Oi settlement on road to Kensu mine from Kashgar, July 17; 4-5 km east of Kensu mine on road to Kashgar from Irkeshtam, July 18; 10 km south-east of Baikurt settlement on old road to Kashgar from Turugart, July 21–1959, Yun. et al).

IIA. Junggar: *Jung. Alat.* (ascent to Kuzyun' pass, Aug. 28, 1908 – B. Fedtsch.), *Tien Shan* (Ketmen' mountain range, at emergence of Sarbushin river from mountains, Aug. 22; 3-4 km beyond Sarbushin settlement, Aug. 23; between Manas and Khorgos rivers, 3 km east of Nyutsyuantsza, Aug. 24; Ketmen' mountain range, descent into Takes river valley, Aug. 26; 42 km from bridge on Kash river on road to Ziekta, Aug. 29; 89 km east of Kul'dzha, Aug. 29–1957, Yun. et al).

General distribution: Jung.-Tarb., Nor. Tien Shan; endemic in Cent. Asia.

120. A. fedtschenkoana Krasch. in Acta Inst. Bot. Ac. Sci. URSS, 1, 3 (1936) 361; Poljak. in Fl. SSSR, 26 (1961) 600, excl. syn. *A. issykkulensis* Poljak.; Filat. in Novit. syst. pl. vasc. 21 (1984) 174; id. 23 (1986) 227; Opred. rast. Sr. Azii [Key to Plants of Mid. Asia] 10 (1993) 573, excl. syn. *A. issykkulensis* Poljak. — *Seriphidium fedtschenkoanum* (Krasch.) Poljak. in Trudy Inst. Bot. Ac. Nauk Kazakhsk. SSR, 11 (1961) 176; Fl. Sin. 76, 2 (1991) 268. — Ic.: Filat. (1986) Plate 4, fig. 11, 12.

Described from Pamir (east. slope of Sarykol' mountain range). Type in St.-Petersburg (LE). Plate VI, fig. 1.

On coastal terraces, pebble beds, loessial slopes of high mountains.

IIIC. Pamir (valley of Tiznaf river, June 8, 1889 – Rob.; Tashkurgan, July 25 – typus!; Kenkol, Togot-Bashi, July 30 – syntypus!; along Kensu river, Aug. 8-1913, Knorr., syntypus!).

General distribution: Cent. Tien Shan; endemic in Cent. Asia.

121. A. fulvella Filat. et Ladyg. in Novit. syst. pl. vasc. 18 (1981) 227; Filat. in Novit. syst. pl. vasc. 21 (1984) 176; id. 23 (1986) 227; Opred. rast. Sr. Azii [Key to Plants of Mid. Asia] 10 (1993) 573. — Ic.: Filat. (1986) Plate 4, fig. 11, 12.

Described from Mid. Asia. Type in St.-Petersburg (LE).

On clayey and clayey-loessial slopes of mountains, forms cereal grass-wormwood associations with shrubby vegetation at 1600–2300 m above sea.

IIA. Junggar: *Tien Shan* (Kul'dzha region, 1877—Fet.; Kul'dzha, valley of Tekes river, Sept. 1876—A. Reg.; Ketmen' mountain range, 3-4 km beyond Sarbushin settlement, Aug. 23; on road to Kyzyl-Kure from Kul'dzha, Aug. 29–1957, Yun., Li et Yuan').

124   General distribution: Nor. and Cent. Tien Shan; endemic in Cent. Asia.

122. A. gobica (Krasch.) Grub. in Novit. syst. pl. vasc. 9 (1979) 296; Grub. Konsp. fl. MNR [Conspectus of Flora of Mongolian People's Republic] (1955) 265; Leonova in Grub. Opred. rast. Mong. [Key to Plants of Mongolia] (1982) 253; Filat. in Novit. syst. pl. vasc. 21 (1984) 166; id. 23 (1986) 234.—*A. mongolorum* Krasch. subsp. *gobica* Krasch. et var. *salsuginosa* Krasch. in Tr. Bot. inst. AN SSSR, 1, 3 (1936) 350; Gub. Konsp. fl. Vneshn. Mong. [Conspectus of Flora of Outer Mongolia] (1996) 98.—*Seriphidium nitrosum* (Web. ex Stechm.) Poljak. var. *gobicum* (Krasch.) Y.R. Ling in Bull. Bot. Res. 8, 3 (1988) 9; Fl. Intramong. ed. 2, 4 (1993) 666. —Ic.: Filat. (1986) Plate 9, fig. 5, 6.

Described from Mongolia (Val. Lakes). Syntypes in Moscow (MW) and St.-Petersburg (LE). Map 8.

In solonchaks, solonetzic meadows, rarely in desert steppes, sand, on columnar solonetzes, Gobi clays, lake solonchaks and coastal pebble beds. Halophyte of desert zone.

IA. Mongolia: *Mong. Alt.* (Bombotu-Khairkhan mountains, in floodplain of Khust river, Oct. 10, 1930—Pob., syntypus!), *East. Mong.* (Choibalsan, Kerulen-gol river stream 20 km north-east of Enger-Shand, Aug. 17, 1949—Yun.), *Depr. Lakes* (Khara-Usu lake, Aug. 13, 1930—Bar.; vicinity of Shargain-Tsagan-Nur lake, Sept. 11, 1930—Pob., syntypus!; nor. bank of Khirgiz-Nur lake, Shitsirgan-bulak area, Aug. 21, 1944—Yun.; Buyantu river basin, Aug. 2, 1971—Grub.; 28 km on Otgon-Ulyasutai road, July 16; 8 km east of Khobdo-Nur lake, July 26–1974, Damba et al), *Val. Lakes* (Tuin-gol river, Sept. 4, 1886 —Pot.; Tuin-gol river, Sept. 6; along Tuin-gol and Tatsin-gol river valleys, Sept. 13–1924, Pavl., syntypus!; Tsagan-gol river near Bain-Gobi somon, Sept. 4; 8 km north of Guchin-Usu somon on road to Arbai-Khere, Sept. 21, 1943, Yun.; Ologoi-Nur lake 10 km south-east of somon, Sept. 1, 1952—Davazhamts), *Gobi Alt.* (Bain-Tukhum area, 20–25 km west of somon, July-Aug. 1933—Simukova; 30 km from Barun-Saikhan mountains, Sept. 25, 1931—Ik.-Gal., syntypus!; from Dundu-Saikhan to Bain-Tukhum, Aug. 28, 1931—Ik.-Gal., syntypus!), *East. Gobi* (16 km north-east of Sain-Shanda on road to Baishintu, Aug. 3, 1940; 165 km south-east of Ulan-Bator to Dalan-Dzadagad, Aug. 1944 —Yun.; Bayan-Dzag, Sept. 9, 1951—Kal.; 20–25 km north—north-east of Sain-Shanda, Aug. 4, 1971—Isach. et Rachk.), *West. Gobi* (Dzakhoi-Dzaram area, Aug. 18; Tsagan-Bogdo area, Aug. 4–1943, Yun.), *Alash. Gobi* (Kobden-Usu area in gobi north of Gashun-Nor lake, Aug. 13, 1886—Pot., syntypus!).

IIA. Junggar: *Zaisan* (Barbagai-Burchum, Sept. 19, 1956—Ching), *Jung. Gobi* (55 km south of Uenchi somon, Boro-Tsondzh area, Aug. 28, 1973—Isach. et Rachk.; Bodonchin-gol floodplain, in downstream, 15 km south-west of Altay settlement, Aug. 1, 1979—Gub.).

General distribution: West. Sib. (Altay), Nor. Mong. (Hang.).

123. **A. gorjaevii** Poljak. in Not. Syst. (Leningrad), 16 (1954) 419; Filat. in Novit. Syst. pl. vasc. 21 (1984) 171; id. 23 (1986) 227. — *A. sublessingiana* auct. non Krasch. ex Poljak.: Fl. SSSR, 26 (1961) 596; Fl. Kazakhst. 9 (1966) 131; Opred. rast. Sr. Azii [Key to Plants of Mid. Asia] 10 (1993) 570. — Ic.: Filat. (1986) Plate 5, fig. 3, 4.

Described from East. Kazakhstan (Jung. Ala Tau). Type in St.-Petersburg (LE).

On steppified slopes of mountains. Dominant species forming steppe cereal grass-wormwood associations at (1200) 1500–2000 m above sea.

IIA. Junggar: *Tarb.* (15 km north of Dachen town, No. 1496, Aug. 12, 1957 — Kuan), *Jung. Gobi* (south. Barbagai, No. 2860, Sept. 8, 1956 — Ching; north of Dzhagistai on road to Chantal, Aug. 7, 1957 — Kuan; Sasheng-ula mountain range, 20 km from Altay somon, July 2, 1973 — Golubk., Tsogt; 27 km south-east of Khairkhan somon on road to Bayan, Aug. 20, 1979 — Grub., Dar. et al; 140 km south-east of Altay somon, Aug. 24, 1981 — Rachk.; west. foothills of Dzhergalan-ul mountain range 100 km from Altay somon, July 26, 1979 — Gub.).

General distribution: Jung.-Tarb., Nor. Tien Shan; endemic in Cent. Asia.

125   Note. Fairly distinct morphological differences (weak pubescence, short (1–3 mm long) leaf lobules, panicle, anthodia generally on well-developed short stalks, and other characteristics) as well as adaptation to steppe associations do not permit identifying this species with the eastern Kazakhstan petrophyte species *A. sublessingiana* Krasch. ex Poljak.

124. **A. gracilescens** Krasch. et Iljin in Animadv. syst. Herb. Univ. Tomsk. 1, 2 (1949) 123; Krasch. in Kryl. Fl. Zap. Sib. 11 (1949) 2785; Grub. Konsp. Fl. MNR [Conspectus of Flora of Mongolian People's Republic] (1955) 265; Poljak. in Fl. SSSR, 26 (1961) 591; Fl. Kazakhst. 9 (1966) 122; Leonova in Grub. Opred. rast. Mong. [Key to Plants of Mongolia] (1982) 253; Filat. in Novit. syst. pl. vasc. 21 (1984) 167; id. 23 (1986) 234; Opred. rast. Sr. Azii [Key to Plants of Mid. Asia] 10 (1993) 567; Gub. Konsp. fl. Vneshn. Mong. [Conspectus of Flora of Outer Mongolia] (1996) 97. — *Seriphidium gracilescens* (Krasch. et Iljin) Poljak. in Trudy Inst. Bot. Ac. Nauk Kazakhsk. SSR, 11 (1961) 175; Fl. Sin. 76, 2 (1991) 274. — Ic.: Fl. Kazakhst. Plate 13, fig. 6; Filat. (1986) Plate 8, fig. 7, 8.

Described from West. Siberia (Kulund steppe). Type in Tomsk (TK). Isotype in St.-Petersburg (LE).

On solonetzic rocky-rubbly slopes of low mountains, trails, in desertified solonetzic steppes, arid beds of seasonal streams. Zonal desert-steppe species forming cereal grass-wormwood associations with shrubs.

IA. Mongolia: *Mong. Alt.* (Gub. l.c.).

IIA. Junggar: *Cis-Alt.* (east of Kran river, Aug. 5, 1876 — Pot.; 30 km east of Burchum on road to Shara-Sume, July 5, 1959 — Yun. et al), *Tarb.* (45 km of Orkhu settlement on Dam river, July 4; 45 km north of Orkhu settlement on road to Altay from Karatai, July

4–1959, Yun. et al), *Jung. Alt.* (Dzhair mountain range, Aug. 3; Urkashar area, 70 km from Toli settlement, Aug. 5; Dzhair mountain range and Urkashar, on road to Chuguchak, Aug. 6; Chuguchak basin, Barlyk mountain range, Aug. 8–1957, Yun. et al), *Jung. Gobi* (Baitak-Bogdo mountain range, nor. trail, 10 km from first conical hillock, Sept. 18, 1948 – Grub.; nor.-west. part of Junggar plain, June 22; 45 km north of Orkhu settlement on road to Altay from Karamai, June 22–1957; nor.-east. part of Junggar plain on way to Guchen from Altay, July 16–1959, Yun. et al; 14 km south-east of Altay settlement on road to Khairkhan, Aug. 12; Uenchiin-gol river valley 4 km before Putsgeniin-gol estuary, Aug. 15; Bulugun river valley 1 km from Ulyastyin-gol estuary, Aug. 17; 28 km south-east of Altay somon centre on road to Khairkhan somon, Aug. 19 –1979, Grub., Dar. et al; Khaldzan-ula mountains, 50 km south-east of Altay settlement, Aug. 26, 1979 – Gub.; 220 km south of Tsogt somon, July 19, 1981 – Rachk.).

General distribution: Fore Balkh., Jung.-Tarb., West. Sib. (Altay).

125. A. heptapotamica Poljak. in Not. Syst. (Leningrad), 18 (1957) 278; Fl. Kazakhst. 9 (1966) 124; Filat. in Novit. syst. pl. vasc. 21 (1984) 173; id. 23 (1986) 227; Opred. rast. Sr. Azii [Key to Plants of Mid. Asia] 10 (1993) 568; Gub. Konsp. fl. Vneshn. Mong. [Conspectus of Flora of Outer Mongolia] (1996) 97.— *A. terrae-albae* Krasch. var. *heptapotamica* (Poljak.) Poljak. in Fl. SSSR, 26 (1961) 563. — *Seriphidium heptapotamicum* (Poljak.) Ling et Y.R. Ling in Fl. Sin. 76, 2 (1991) 276. — Ic.: Filat. (1986) Plate 5, fig. 9, 10.

Described from East. Kazakhstan (between Kungei Ala Tau mountain range and Turaigyr). Type in St.-Petersburg (LE). Map 8.

On rubbly, clayey, rubbly-loessial slopes of low mountains, pebble beds of seasonal streams, exposed red clays on rubbly-clayey soils in intermontane plains. Dominant desert steppe and desert species at 1000–1300 m above sea.

126   IA. Mongolia: *Mong. Alt.* (valley of Bulgan river, beyond Bulgan somon, in lower and middle streams, Aug. 7, 1984 – Kam. et Dar.).

IIA. Junggar: *Jung. Alt.* (in Toli town region, Aug. 3, 1951 – Kuan; valley of Borotala river at crossing near Khundelyun settlement, Aug. 18, 1957 – Yun. et al), *Tien Shan* (Kul'dzha region, July 24, 1877 – A. Reg.; Nintszyakhe, Sept. 30, 1956 – Ching; offshoot of Bogdo-ul, June 2; between Manas and Khorgos rivers, July 24; trails of Ketmen' mountain range, Aug. 21; Tekes river, 5–6 km north-east of Aksu settlement on road to Kalmak-Kure, Aug. 25; left flank of Kunges river valley, Aug. 28; left bank of Kunges river 8–10 km north-west of state farm of 10th regiment on road to pass, Aug. 29; 89 km east of Kul'dzha, on right bank of Ili-Kunges river valley, Aug. 29; Borokhoro mountain range trails, Sept. 1–1957, Yun. et al; nor. foothills of Tien Shan, Aug. 30; 55 km west of Urumchi, Oct. 2; 17 km east of Muleikhe, Oct. 3–1959, Petr.), *Jung. Gobi* (Kuitun river, June 29; on road to Urumchi from Guchen, July 17; on Manas-Shikho road, Aug. 2, 1959 –Yun. et al; 17 km west of Chantsza town, Sept. 17; 30 km south-east of Gan'khetsza, Sept. 23–1957, Kuan; 28 km south-east of Altay somon at Bodonchiin-gol on road to Khairkhan, rocky slopes of conical hillocks, desert, Aug. 19; 15 km east of Tsargin post north of Takhiin-Shara-nuru, in gorge, Aug. 20–1979, Grub., Muld. et Dar.; 40 km south – south-east of Bulgan somon, Palaeogene chalk formations, Aug. 19; 75 km south of Bulgan somon, desert, 1300 m, Aug. 19; 17 km east of Khairkhan ruins, in pea shrub scrubs, Aug. 23– 1984, Rachk. et Dar.; right bank of Bulgan river near Bulgan somon, July 18; south. Argalant-ul mountain, hummocky area, July 31; Dzygin-Ulan-ula low mountains, Aug. 18 –1984, Kam. et Dar.).

General distribution: Jung.-Tarb., Nor. Tien Shan, endemic in Cent. Asia.

Note. Exhibits extremely close affinity with *A. sublessingiana* Krasch. ex Poljak. and not with *A. terrae-albae* Krasch. as P.P. Poljakov did in the cited reference. In our view, this is one more ecological-geographic race separated from northern *A. sublessingiana* and differing from it in very short and soft, deciduous leaves or leaves wilting in summer, pyramidal panicle, much smaller anthodia and phyllary with distinctly manifest golden border.

126. A. issykkulensis Poljak. in Not. syst. (Leningrad), 17 (1955) 415; Fl. Kirgiz. 11 (1965) 193; Filat. in Novit. syst. pl. vasc. 21 (1984) 174; id. 23 (1986) 227. – *A. fedtschenkoana* auct. non Krasch.: Poljak. in Fl. SSSR, 26 (1961) 600; Opred. rast. Sr. Azii [Key to Plants of Mid. Asia] 10 (1993) 573. – *Seriphidium issykkulense* (Poljak.) Poljak. in Trudy Inst. Bot. Ac. Nauk. Kazakhsk. SSR, 11 (1961) 173; Fl. Sin. 76, 2 (1991) 269. – Ic.: Filat. (1986) Plate 5, fig. 7, 8.

Described from Tien Shan (Issyk-Kul' lake basin). Type in St.-Petersburg (LE). Map 8.

On rubbly-clayey and rocky slopes, in beds of arid gullies, on pebble beds. Dominant species, forms wormwood-bean caper, wormwood-saltwort, wormwood-*Reaumuria* cenoses at 1200–1800 m above sea.

IB. Kashgar: *Nor.* (Muzart river cone around Alty-Gumbez settlement on road to Charchi, Sept. 6; 4–5 km south of Kzyl-Bulak settlement, on road to Avat, Sept. 13; east of Bai town on road to Aksu from Kucha, Sept. 21; 24–25 km west of Bai town on road to Aksu, Sept. 21; 2 km east of Ushak-tal settlement on road to Karashar, Sept. 28–1957; 20–25 km north-west of Kucha settlement, Aug. 31; Taushkan-dar'ya river valley 30–49 km south-west of Uch-Turfan Sept. 17–1958, Yun., Li et Yuan').

General distribution: Cent. Tien Shan; endemic in Cent. Asia.

127   Note. Distinct species with stable characteristics affiliated to *A. heptapotamica* Poljak. and succeeding the latter in mid-mountain deserts. P.P. Poljakov (l.c.) erroneously (probably for want of adequate herbarium material) regarded it as a synonym of *A. fedtschenkoana* Krasch.

127. A. juncea Kar. et Kir. in Bull. Soc. natur. Moscou, 15 (1842) 383; Krasch. in Kryl. Fl. Zap. Sib. 11 (1949) 2788; Poljak. in Fl. SSSR, 26 (1961) 629; Fl. Kirgiz. 11 (1965) 201; Fl. Kazakhst. 9 (1966) 116; Filat. in Novit. syst. pl. vasc. 21 (1984) 165; id. 23 (1986) 219; Opred. rast. Sr. Azii [Key to Plants of Mid. Asia] 10 (1993) 586. – *Seriphidium junceum* (Kar. et Kir.) Poljak. in Trudy Inst. Bot. Ac. Nauk. Kazakhsk. SSR, 11 (1961) 175; Fl. Sin. 76, 2 (1991) 288. – Ic.: Filat. (1986) Plate 1, fig. 9, 10.

Described from East. Kazakhstan (Ayaguz river basin). Type in Moscow (MW).

On rocky and stony slopes of low mountains and rubbly-sandy alluvium in dry beds of seasonal streams. Desert-steppe petrophyte species.

IIA. Junggar: *Tien Shan* (38 km east of bridge of Kash river on road to Ziekta, right flank of Ili-Kunges valley, Aug. 29; same site, 42 km east of bridge on Kash river, Aug.

29; same site, 89 km east of Kul'dzha, Aug. 29–1957, Yun. et Yuan'), *Jung. Gobi* (Karamai- Shchikhedzy, No. 3397, Sept. 24, 1956—Ching; 60 km north-east of Tsitai town, No. 2323, Sept. 25, 1957—Kuan; left bank of Chernyi Irtysh river, 35–40 km south-east of Shipata settlement in Irtysh-Urungu interfluvine region, July 8, 1959—Yun. et Yuan').

General distribution: Fore Balkh., Jung.-Tarb., Nor. Tien Shan, Cent. Tien Shan, Kopetdag.

128. A. kaschgarica Krasch. in Acta Inst. Bot. Ac. Sci. URSS, 1, 3 (1936) 350; Poljak. in Fl. SSSR, 26 (1961) 599; Fl. Kazakhst. 9 (1966) 128; Filat. in Novit. syst. pl. vasc. 21 (1984) 166; id. 23 (1986) 227. — *Seriphidium kaschgaricum* (Krasch.) Poljak. in Trudy Inst. Bot. Ac. Nauk. Kazakhsk. SSR, 11 (1961) 175; Fl. Sin. 76, 2 (1991) 267.

Described from East. Tien Shan (Daban'chen area). Type in St.-Petersburg (LE).

On rubbly-clayey slopes of low mountains, riverine pebble beds, loessial alluvia, on arid river beds, in saxaul forests.

IIA. Junggar: *Tien Shan* (Koksu, Aug. 2, 1898—Klem.; Daban'-chen area, Sept. 4, 1929 —Pop., typus!; south. Urumchi settlement, July 19, 1956—Ching; 89 km from Kul'dzha, valley of Ili-Kunges river, Aug. 29; same site, on road to Ziekta from Ili, Aug. 29–1957, Yun., Li et Yuan'), *Jung. Gobi* (south. Shara-Sume, No. 2847, Sept. 8; south. Barbagai, No. 2861, Sept. 8–1956, Ching; San'da-okhedza, July 4; 5 km west on road to Kuitun, July 30 –1957, Kuan; 3 km toward east of San'daokhedza on road to Manas from Tushantsza, July 4; 3 km east of bridge on Yantszykhai on road to Shito from Manas, July 7; 10 km north-east on road to Karaganda pass, Aug. 14–1957, Yun., Li et Yuan').

General distribution: Jung.-Tarb.; endemic in Cent. Tien Shan.

Note. This species was described by I.M. Krascheninnikov from the collections of M.G. Popov from Junggar (East. Tien Shan). M.G. Popov erroneously hand-wrote "Kaschgaria" on the label of type.

128   129. A. mongolorum Krasch. in Acta Inst. Bot. Ac. Sci. URSS, 1, 3 (1936) 350; Krasch. in Kryl. Fl. Zap. Sib. 11 (1949) 2782; Filat. in Novit. syst. pl. vasc. 23 (1986) 236; Gub. Konsp. fl. Vneshn. Mong. [Conspectus of Flora of Outer Mongolia] (1996) 98. — *A. nitrosa* auct. non Web. ex Stechm.: Poljak. in Fl. SSSR, 26 (1961) 580. — *Seriphidium mongolorum* (Krasch.) Ling et Y.R. Ling in Bull. Bot. Res. 8, 3 (1988) 115; Fl. Sin. 76, 2 (1991) 265; Fl. Intramong. ed. 2, 4 (1993) 664. — Ic.: Filat (1986) Plate 9, fig. 9, 10.

Described from Qaidam (Orogyn-gol river). Type in St.-Petersburg (LE). Map 8.

On loessial, rubbly-clayey slopes of low mountains, in arid loamy and pebbly beds of gorges, rarely on standing fallow lands.

IA. Mongolia: *East. Gobi* (Gub. l.c.), *Gobi Alt.* (south. slope of Dzuisaikhan mountain range, 45 km north-east, Sept. 15, 1951—Kal.; Naran-Bulak area, 1500 m above sea, Aug. 12, 1979—Gub.), *West. Gobi* (70 km south of Bugot somon, nor. slope, Aug. 24, 1984— Rachk., Buyan-Orshikh et Dar.), *Alash Gobi* (valley of Edzin-gol river, Bukhan-Khub area, July 13, 1926—Glag.).

IC. Qaidam: *Plain* (nor. slope of Burkhan-Budda mountain range, Aug. 16, 1884 – Przew.; 91 km east of Golmo settlement, Oct. 13, 1959 – Petr.), *Mount.* (valley of Orogyn-gol river, Aug. 15, 1879 – Przew., typus!).

IIA. Junggar: *Jung. Gobi* (arid bed of Barlagiin-gol on road to Bodonchiin-Baishing, Sept. 9, 1948 – Grub.; Paotai forest industry on Manas river, No. 3850, Oct. 10, 1956 – Ching; forest industry on Manas river in Syaeda region, June 13; Tsitai-Meiyao, No. 2346, Sept. 16-1957, Kuan).

General distribution: endemic in Cent. Asia.

130. A. nitrosa Web. ex Stechm. Artemis. (1775) 24; Krasch. in Kryl. Fl. Zap. Sib. 11 (1949) 2781; Poljak. in Fl. SSSR, 26 (1961) 580; Fl. Kazakhst. 9 (1966) 126; Filat. in Novit. syst. pl. vasc. 21 (1984) 166; id. 23 (1986) 234; Gub. Konsp. fl. Vneshn. Mong. [Conspectus of Flora of Outer Mongolia] (1996) 98. – *A. maritima* L. var. *dahurica* Turcz. in Fl. baic.-daur. 2 (1845) 56. – *A. dahurica* (Turcz.) Poljak. in Grub. Konsp. fl. MNR [Conspectus of Flora of Mongolian People's Republic] (1955) 264. – *Seriphidium nitrosum* (Web. ex Stechm.) Poljak. in Trudy Inst. Bot. Ac. Nauk Kazakhsk. SSR, 11 (1961) 172; Fl. Sin. 76, 2 (1991) 261; Fl. Intramong. ed. 2, 4 (1993) 666. – Ic.: Filat. (1986) Plate 9, Fig. 7, 8.

Described from East. Siberia (upper course of Chulym river). Type in Moscow (MW).

On solonetzic steppe meadows, along banks of saline lakes, in floodplains of rivers, Halophyte of forest-steppe and steppe zones.

IA. Mongolia: *Cent. Khalkha* (165 km south-east of Ulan-Bator on road to Sain-Shanda, Aug. 21, 1949 – Yun.), *East. Mong.* (Kerulen-gol river current, 20–25 km north-east of Enger-Shanda, Aug. 17; 35-40 km north-east of Enger-Shanda, Aug. 17-1949, Yun.; 100 km south-east of Choibalsan town, Aug. 17, 1954 – Dashnyam).

General distribution: West. Sib. (Altay), East. Sib. (Sayans), Nor. Mong. (Hang., Mong.-Daur.).

131. A. saissanica (Krasch.) Filat. in Tr. Inst. bot. AN Kaz. SSR, 15 (1963) 234; Filat. in Novosti sist. rast. Inst. bot. AN Kazakhst. 3 (1965) 43; Fl. Kazakhst. 9 (1966); Filat. in Novit. syst. pl. vasc. 19 (1982) 177; id. 23 (1986) 231; Opred. rast. Sr. Azii [Key to Plants of Mid. Asia] 10 (1993) 573. – *A. mongolorum* Krasch. subsp. *saissanica* Krasch. in Kryl. Fl. Zap. Sib. 11 (1949) 2783; Gub. Konsp. Fl. Vneshn. Mong. [Conspectus of Flora of Outer Mongolia] (1996) 98. – Ic.: Filat. (1986) Plate 5, fig. 9, 10.

Described from East. Kazakhstan (Zaisan basin). Lectotype in St.-Petersburg (LE).

In meadowy solonchaks, solonchak chee grass thickets, along banks of saline lakes, along rocky slopes of low mountains, arid beds of rivers.

IA. Mongolia: *Mong. Alt., Depr. Lakes, Val. Lakes* (Gub. l.c.).

IIA. Junggar: *Cis-Alt.* (Chernyi Irtysh river basin on road to Shara Sume, July 7, 1959 – Yun., Li, Yuan' et al.), *Jung. Alt.* (Dzhair mountain range, 24 km north-east of Toli settlement on road to Temirtam, Aug. 5; on road to Chugurchak, Aug. 7; 12 km south-east of Kurte settlement on road to Toli from Durbul'dzhin, Aug. 8–1957, Yun., Li,

Yuan', *Jung. Gobi* (10 km south of Shikho settlement, Chidachu oasis, Aug. 10, 1947 —
Shumakov; Ulyungur lake, July 9, 1959 — Yun., Li et Yuan').

General distribution: Jung.-Tarb., Nor. Tien Shan; endemic in Cent. Asia.

129 132. **A. santolina** Schrenk in Bull. Ac. Sci. (St.-Petersb.) 3, 7 (1845) 106;
Poljak. in Fl. SSSR, 26 (1961) 625; Fl. Kazakhst. 9 (1966) 115; Filat. in
Novit. syst. pl. vasc. 21 (1984) 165; id. 23 (1986) 231; id. Opred. rast. Sr.
Azii [Key to Plants of Mid. Asia] 10 (1993) 586. — *Seriphidium santolinum*
(Schrenk) Poljak. In Trudy Inst. Bot. Ac. Nauk Kazakhsk. SSR, 11 (1961)
173; Fl. Sin. 76, 2 (1991) 280. — Ic.: Filat. (1986) Plate 7, fig. 7, 8.

Described from East. Kazakhst. (Ili river). Type in St.-Petersburg (LE).

On hummocky and dune sand.

IIA. Junggar: *Jung. Gobi* (left bank of Manas river 20 km north-west of Savan
settlement, June 10; 20–25 km north of Da-Myao on road to Mo-Savan, July 10; south.
fringe of Dzosotyn sand, July 11–1957; east. part of Dzosotyn Elisun sand 88 km north
of Guchen, July 16, 180–185 km south of Ertai settlement on road to Guchen from Altay,
July 16–1959, Yun., Li et Yuan').

General distribution: Aralo-Casp., Fore Balkh., Fore Asia, Mid. Asia (plain and
mountainous).

133. **A. schischkinii** Krasch. in Animadv. Syst. Herb. Univ. Tomsk, 1,
2 (1949) 2; Krasch. in Kryl. Fl. Zap. Sib. 11 (1949) 2786; Grub. Konsp. fl.
MNR [Conspectus of Flora of Mongolian People's Republic] (1955) 268;
Leonova in Grub. Opred. rast. Mong. [Key to Plants of Mongolia] (1982)
253; Filat. in Novit. syst. pl. vasc. 23 (1986) 234; Gub. Konsp. fl. Vneshn.
Mong. [Conspectus of flora of Outer Mongolia] (1996) 99. — *A. nitrosa*
auct. non Web. ex Stechm.: Poljak. in Fl. SSSR, 26 (1961) 580.

Described from Altay (Chui steppe). Type in St.-Petersburg (LE).

In mountain desert steppes, desert steppes with taro (*Colocasia*), on
rocky slopes, rocks, pebble beds of rivers.

IA. Mongolia: *Khobd.* (Gub. l.c.); *Mong. Alt.* (within Ogvoldzo daban, Aug. 6; along
bank of Bidzha river, Aug. 6–1896, Klem.; Khara-Adzarga mountains, valley of Sakhir-
sala river, Aug. 24, 1930 — Pob.; upper course of Bodonchi river, Sept. 10; valley of
Uinchi river, near Ulyastu-Tokhoi, Sept. 13; Tergin-gol, Sept. 14–1930, Bar.; valley
between Beger-Nur and Shargain-gol, Aug. 23; Khalyun area, near Khalyun, Aug. 24–
1943, Yun.; Tamchi-daba pass, Sept. 8; Bodonchiin-gol at exit of gorge, 45 km from
Bodonchiin-Khere, Sept. 25–1948; Uenchiin-gol river basin, right bank of Kharchin-gol,
3 km from estuary, Aug. 18–1979, Grub., Dar. et al), *Depr. Lakes* (between Khara-Bura
river and Dzemen well, Sept. 1; between Kedryn'-gol river and Ubsa-Nor lake, Sept. 4–
1985, Klem.; between Gol-Ikhe and Tszak-obo, bank of Shargain-gol river, Sept. 6; at
Borin-Khotok Well, Oct. 3–1930, Pob.).

IIA. Junggar: *Tarb.* (Semistei mountain range, 15 km west of Kosh-Tologoi, June 22,
1957 — Yun. et al.), *Jung. Gobi* (south of Bain-Tsagan-ula, June 26, 1944; 23 km west of
Birzhi-gol, on road to Bulugun, July 18; Barlagin-Ubchugin-gol interfluvine region and
nor. extremity of Argalantu-ul, July 18–1947, Yun.; 6 km from Bulugun somon on road
5 to Uienchi somon, Sept. 22, 1948 — Grub.).

General distribution: West. Sib. (Altay).

134. A. schrenkiana Ledeb. Fl. Ross. 2, 2 (1845) 575; Krasch. in Kryl. Fl. Zap. Sib. 11 (1949) 2783; Grub. Konsp. Fl. MNR [Conspectus of Flora of Mongolian People's Republic] (1955) 582; Fl. Kirgiz. 11 (1965) 180; Fl. Kazakhst. 9 (1966) 127; Leonova in Grub. Opred. rast. Mong. [Key to Plants of Mongolia] (1982) 253; Filat. in Novit. syst. pl. vasc. 21 (1984) 170; id. 23 (1986) 234; Opred. rast. Sr. Azii. [Key to Plants of Mid. Asia] 10 (1993) 565; Gub. Konsp. fl. Vneshn. Mong. [Conspectus of Flora of Outer Mongolia] (1966) 99. — Ic.: Filat. (1986) Plate 8, fig. 7, 8.

Described from East. Kazakhstan (Tarbagatai mountain range). Type in St.-Petersburg (LE).

On highly saline soils, meadowy and puffed solonchaks, along banks of saline lakes, on river terraces, in wormwood-taro (with Nanophyton) and steppified deserts.

IA. Mongolia: *Khobd., Depr. Lakes* (Gub. l.c.).

130   IIA. Junggar: *Cis-Alt.* (left bank of Chernyi Irtysh river, June 8, 1914 — Schischk.; Barbagatai-Burchum, Sept. 11, 1956 — Ching), *Jung. Alt.* (in foothills of Dzhair mountain range, Aug. 5, 1951 — Mois.), *Jung. Gobi* (Dzhirgalantu-gol, left tributary of Bulugun, Sept. 1930 — Bar.; Bulgan, fringe of Ingiin-us spring, Aug. 20–1981, Rachk.; 80 km south-west of Bulugun settlement, Aug. 15, 1982 — Gub.).

General distribution: Fore Balkh., Jung.-Tarb., West. Sib. (Altay), Nor. Mong. (Hang.).

135. A. skorniakovii Winkl. in Acta Horti Petrop. 2, 2 (1892) 331; Fliat. in Novit. syst. pl. vasc. 23 (1986) 236. — *A. lehmanniana* auct. non Bge.: Poljak. in Fl. SSSR, 26 (1961) 622. — Ic.: Filat. (1986) Plate 10, fig. 1, 2.

Described from Pamir (Kara-Kul' lake basin). Type in St.-Petersburg (LE).

On clayey, loessial-clayey mountain slopes, sand alluvium, on moraines, Dominant species of alpine desert; forms plant associations 3600–4000 m above sea.

IB. Kashgar: *Nor.* (Declivites australis jugi montium Tianschan, Kukurtuk-tal, June 22–July 2, 1903 — Merzb.), *West.* (valley of Sulu-Sakhal river 25 km from Irkeshtam, July 26, 1935 — Olsuf'ev; 25–27 km above Kyude settlement in upper courses of Tiznaf river, June 1, 1959 — Yun.), *South.* (Keriya river basin 4 km south and beyond Polur settlement, May 11; same site, 0.5 km west of Polur settlement, May 11; same site, 10 km north of Polur settlement on road of Keriya, May 13; Chira river basin (right tributary of Sary-Bulak river) 15–17 km south of Uku settlement, May 19–1959, Yun., Yuan').

IIIC. Pamir: *Pamir* (upper courses of Lanet river, July 20, 1942 — Serp.; east. slope of Sarykol' mountain range, 4–5 km west of Tashkurgan town, June 13, 1959 — Yun., Yuan').

General distribution: East. Pamir; endemic in Cent. Asia.

136. A. subchrysolepis Filat. in Novit. syst. pl. vasc. 18 (1981) 224; ibid, 21 (1984) 172; ibid, 23 (1986) 227; Opred. rast. Sr. Azii [Key to Plants of Mid. Asia] 10 (1993) 571; Gub. Konsp. fl. Vneshn. Mong. [Conspectus of Flora of Outer Mongolia] (1996) 99. — Ic.: Filat. (1986) Plate 5, fig. 5, 6.

Described from East. Kazakhstan (Junggar gate). Type in St.-Petersburg (LE). Plate VI, fig. 3. Map 8.

On rubbly-rocky slopes of low mountains, rocky deserts, saxaul forests. Dominant species of desert hammada (rocky desert) type.

IA. Mongolia: *Mong. Alt.* (near Khalyun settlement, July 16; 30 km north-west of Bulgan settlement, Aug. 16–1979, Gub.; valley of Ulyasutain-gol river, upper portion of valley, July 13; basin of Bulugun river, in midcourse, Aug. 24–1984, Kam. et Dar.).

IIA. Junggar: *Cis-Alt.* (valley of Chingil river, July 29, 1906–Sap.; Daban-Tsinkhei, Aug. 8; in Fuyun' town region, No. 2797, Aug. 20–1956, Ching), *Tarb.* (Dzhair-Barlyk-Manas, Sept. 6, 1947–Shumakov; 43 km from Durbul'dzhin south of Toli, July 8; saddle-shaped trough between Dzhair mountain range and Urkashar 3–4 km north of Temirtam settlement, Aug. 4; 6–7 km north-west of Temirtam settlement, Aug. 6; 43 km from Durbul'dzhin southward of Toli, Aug. 8–1957, Yun., Li, Yuan' et al), *Jung. Alt.* (Dzhair mountain range, 10 km west of Aktam settlement, Aug. 3; 14–15 km east of Temirtam settlement on road to Chimpaza, Aug. 6; 30 km east of Borotal settlement on road to Ebi-nur, Aug. 13; Maili mountain range, 20–25 km north-east of Junggar gate mateorological station toward Karaganda pass, Aug. 14–1957, Yun. et al), *Tien Shan* (nor. slope of Mertzbacher mountain range in Santai region, Sept. 5, 1952–Moiseenko; Nintszyakhe, Sept. 30; Shichan foothills, Oct. 1–1956, Ching; on Chapchal-Dzhagastai road, Aug. 6, 1957 –Kuan), *Jung. Gobi* (west. Tsigai, July 23; from Burchum town 80 km south, Aug. 6; 120 km north of Burchum, Sept. 14; Shikhetsza, Oct. 4; Kuitun-Dushan'tszy, Oct. 24–1956, Ching; 18 km west of Chantsza town, Sept. 17; 15 km east of Sebikou, Sept. 24; 49 km north-east of Tsitai town, Sept. 25–1957, Kuan; 37 km north-east of Orkhu settlement on road to Kosh-Tologoi from Karamai, June 22; right bank of Kuitun, south of Tushantsza oil pipeline, June 29; 24 km west of Shikhetsza town on road to Santokhodze settlement, July 3; 42 km north-east of Orkhu settlement on Kosh-Tologoi, July 22; 14–15 km east of Temirtam on road to Chimpatsza, Aug. 6; bank of Ebi-nur lake, Aug. 13; 30 km east of Borotal settlement on road to Ebi-nur lake, Aug. 13; 15 km north-west of Urumchi, Sept. 21; 43 km north-east of Urumchi, Oct. 4; 97 km east of Muleikhe, Oct. 5; 2 km from bridge on Epte on road to Shikho from Dzin'kho, Oct. 21–1957; 15 km south of Ertai on road to Guchin from Altay, July 16, 1959–Yun., Li et Yuan').

General distribution: Jung.-Tarb.; endemic in Cent. Asia.

137. A. sublessingiana Krasch. ex Poljak. in Not. syst. (Leningrad) 16 (1954) 395; Krasch. in Kryl. Fl. Zap. Sib. 11 (1949) 2788; Poljak. in Fl. SSSR, 26 (1961) 596, excl. syn. *A. gorjaevii* Poljak. and *A. polysticha* Poljak.; Fl. Kirgiz. 11 (1965) 192; Fl. Kazakhst. 9 (1966) 131, excl. syn. *A. gorjaevii* Poljak. and *A. polysticha* Poljak.; Leonova in Grub. Opred. rast. Mong. [Key to Plants of Mongolia] (1982) 253; Filat. in Novit. syst. pl. vasc. 23 (1986) 227; Opred. rast. Sr. Azii [Key to Plants of Mid. Asia] 10 (1993) 570; Gub. Konsp. fl. Vneshn. Mong. [Conspectus of Flora of Outer Mongolia] (1996) 98.—*Seriphidium sublessingianum* (Krasch. ex Poljak.) Poljak. in Trudy Inst. Bot. Ac. Nauk Kazakhsk. SSR, 11 (1961) 174; Fl. Sin. 76, 2 (1991) 277. —Ic.: Filat (1986) Plate 5, fig. 5, 6.

Described from East. Kazakhstan (Romanovka settlement). Type in St.-Petersburg (LE).

On rubbly and rocky slopes of low mountains and conical hillocks. East. Kazakhstan species dominating in steppified deserts.

IA. Mongolia: *Gobi Alt.* (Ikhe-Bogdo-ula, July 24, 1973 – Isach. et Rachk.).

IIA. Junggar: *Cis-Alt.* (south. Shara-Sume, Aug. 9; Chingil, No. 1672, Aug. 11; around Shara-Sume, Sept. 4–1956, Ching; 30 km west of Emel' town, No. 1472, Aug. 10, 1957 – Kuan; 28–30 km south of Shara-Sume, on road to Shipota crossing on Chernyi Irtish, July 7; Chernyi Irtysh and Urumchi interfluvine region, 35–45 km south-east of Shipota settlement, July 8–1950; 35 km south of Kok-Togoi on road to Ertai, July 14, 1959 – Yun., Yuan' et al), *Jung. Gobi* (Khubchiin-nuru mountain range, west of Adzhi-Bogdo, Aug. 3; 10 km north of Bidzhiin-gol, Aug. 17–1947, Yun.; Baitak-Bogdo mountain range, July 27, 1979 – Gub.; right bank of Bulgain-gol river, low mountains in Dzun-Khadz-ul town, July 12, 1984 – Kam. et Dar.).

General distribution: Fore Balkh., Jung.-Tarb.; endemic in Cent. Asia.

138. A. terrae-albae Krasch. in Mat. Kompl. eksp. issl. 26 (1930) 269; Krasch. in Kryl. Fl. Zap. Sib. 11 (1949) 2787; Grub. Konsp. fl. MNR [Conspectus of Flora of Mongolian People's Republic] (1955) 269; Poljak in Fl. SSSR, 26 (1961) 592, excl. syn. A. suaveolens Poljak. and A. heptapotamica Poljak.; Fl. Kirgiz. 11 (1965) 188; Fl. Kazakhst. 9 (1966) 120, excl. syn. A. suaveolens Poljak.; Leonova in Grub. Opred. rast. Mong. [Key to Plants of Mongolia] (1982) 253; Filat. in Novit. syst. pl. vasc. 21 (1984) 169; ibid, 23 (1986) 234; Opred. rast. Sr. Azii [Key to Plants of Mid. Asia] 10 (1993) 568; Gub. Konsp. fl. Vneshn. Mong. [Conspectus of Flora of Outer Mongolia] (1996) 98. – Seriphidium terrae-albae (Krasch.) Poljak. in Trudy Inst. Bot. Ac. Nauk Kazakhsk. SSR, 11 (1961) 175; Fl. Sin. 76, 2 (1991) 275. – Ic.: Filat. (1986) Plate 8, fig. 9, 10.

Described from Cent. Kazakhstan (lower courses of Sary-su river). Type in St.-Petersburg (LE).

In wormwood-taro rubbly-clayey, rubbly and sandy deserts, on floors of arid gorges, along takyrs (clay-surfaced deserts), on standing fallow land. North Turan species, dominating in deserts zone.

IA. Mongolia: *Mong. Alt.* (2–3 km south of Tamchi lake, July 17, 1947 – Yun.).

IIA. Junggar: *Tarb.* (valley of Emel' river, Aug. 8, 1957 – Yun. et al), *Jung. Alt.* (Urkashar area, Aug. 5; 12 km south-east of Kurte on road to Toli from Durbul'dzhin, Aug. 8–1957, Yun. et al), *Jung. Gobi* (in Dzhirgalantu-gol river valley, left tributary of Bulugan, Sept. 16; lower course of Uenchi, Sept. 21–1930, Bar.; on Shikho-Karamai road, June 21; 54 km north-west of Padai on road to Chugai, July 17; 20–25 km north of Da-Myao, July 17–1957; Chernyi Irtysh and Urumchi rivers interfluvine region, July 8; on road to Guchen from Altay, July 16–1959, Yun. et al.; 15 km west of Uenchi somon, Aug. 3; 25 km south-west of Bulgan somon, Aug. 6; 120 km east – south-east of Bulgan somon, Aug. 14; 20 km north – north-west of ruins in Khairkhan settlement on road to Bulgan somon, Aug. 23; 45 km east – south-east of Altay somon, Aug. 24–1977, Isach. et Rachk.; 30 km south of Bulgan settlement, July 26; 15 km north of Baitag-Bogdo post, July 27; south of Bulgan settlement, Aug. 26–1979, Gub.), *Dzhark.* (west of Suidun settlement, Aug. 31, 1957 – Yun. et al).

General distribution: Aralo-Casp., Fore Balkh., Jung.-Tarb.; endemic in Cent. Asia.

Plate I

1 — *Achillea acuminata* (Ledeb.) Sch. Bip.; 2 — *A. ptarmicoides* Maxim.;
3 — *Chrysanthemum zawadskii* Herbich; 4 — *Ch. sinuatum* Ledeb.

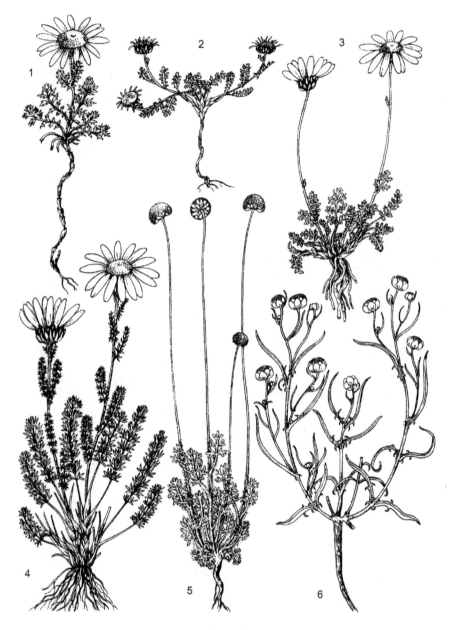

Plate II

1 – *Waldheimia tomentosa* (Decne.) Rgl.; 2 – *W. stoliczkae* (Clarke) Ostenf.;
3 – *Pyrethrum kaschgaricum* Krasch.; 4 – *P. pulchrum* Ledeb.; 5 – *Cancrinia
tianschanica* (Krasch.) Tzvel.; 6 – *Stipnolepis centiflora* (Maxim.) Krasch.

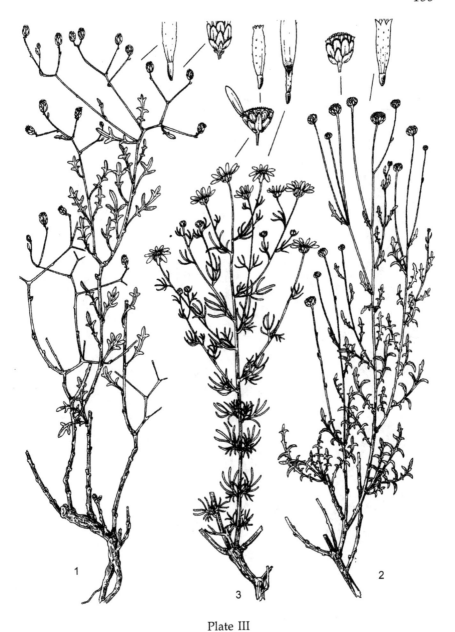

Plate III

1 — *Poljakovia kaschgarica* (Krasch.) Grub. et Filat.; 2 — *P. falcatolobata* (Krasch.)
Grub. et Filat.; 3 — *Brachanthemum nanschanicum* Krasch.

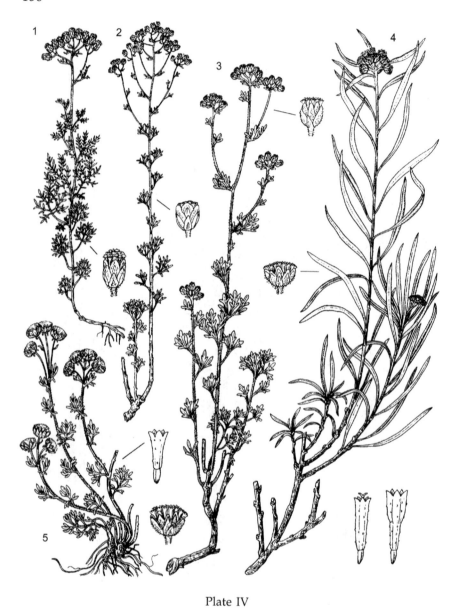

Plate IV

1— *Ajania roborowskii* Muld.; 2— *A. gracilis* (Hook. f. et. Thoms.) Poljak. ex
Tzvel.; 3— *A. grubovii* Muld.; 4— *Phaeostigma salicifolium* (Mattf.) Muld.;
5— *Hippolytia herderi* (Rgl. et Schmalh.) Poljak.

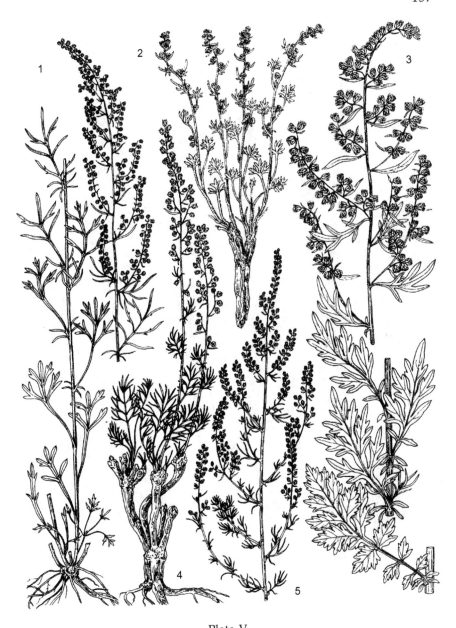

Plate V

1 — *Artemisia manshurica* (Kom.) Kom.; 2 — *A. duthreuil-de-rhinsii* Krasch.;
3 — *A. tangutica* Pamp; 4 — *A. xylorhiza* Krasch. ex Filat.; 5 — *A. ordosica* Krasch.

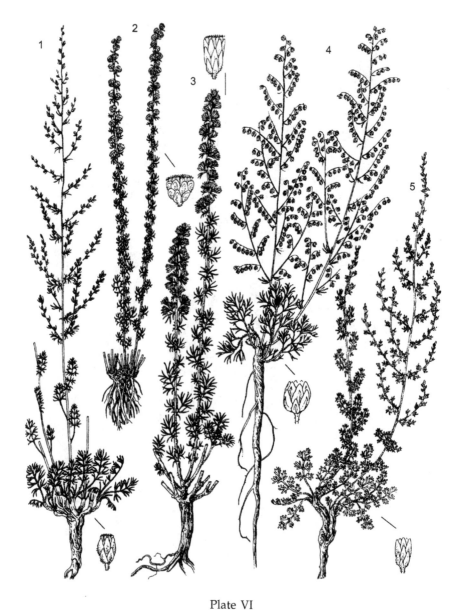

Plate VI

1 — *Artemisia fedtschenkoana* Krasch.; 2 — *A. dalai-lamae* Krasch.; 3 — *A. subchrysolepis* Filat.; 4 — *A. pewzowii* Winkl.; 5 — *A. assurgens* Filat.

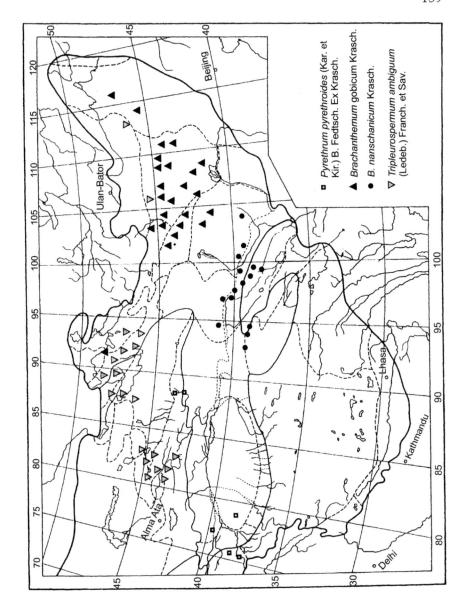

Map 1

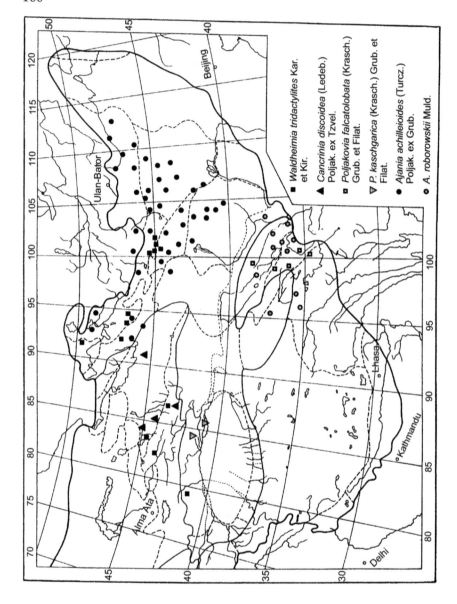

Map 2

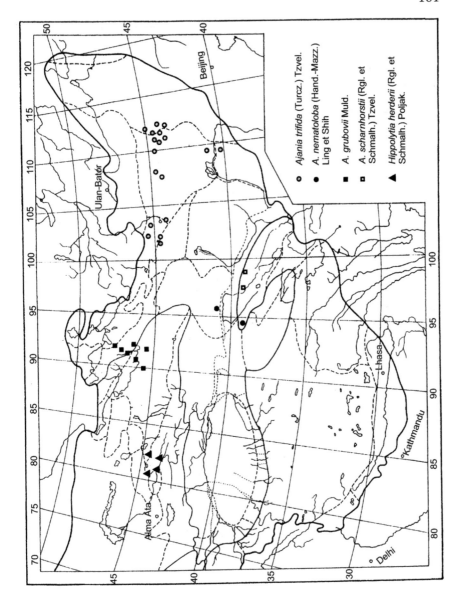

Map 3

162

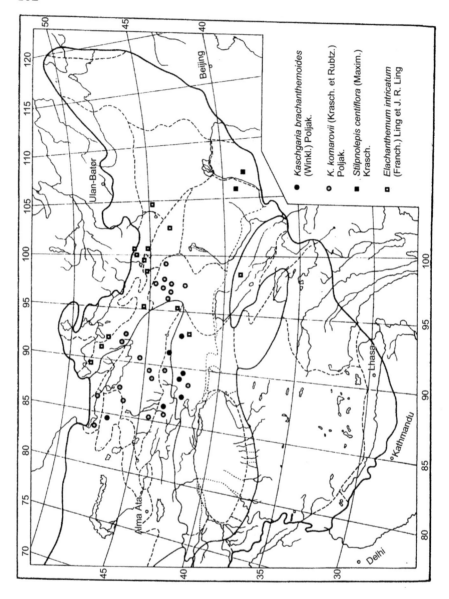

Map 4

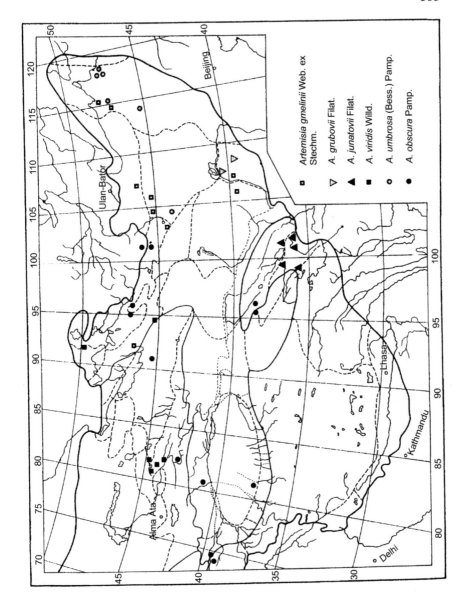

Map 5

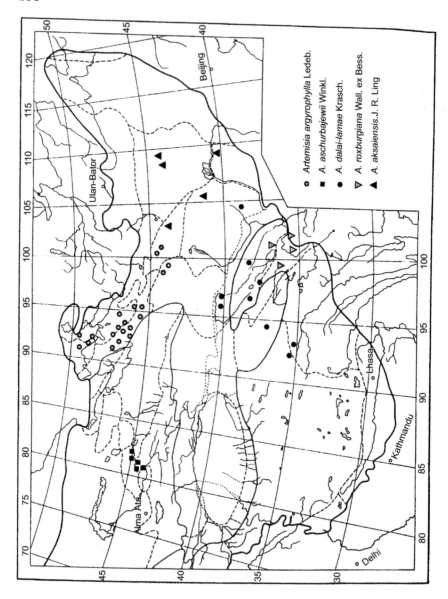

Map 6

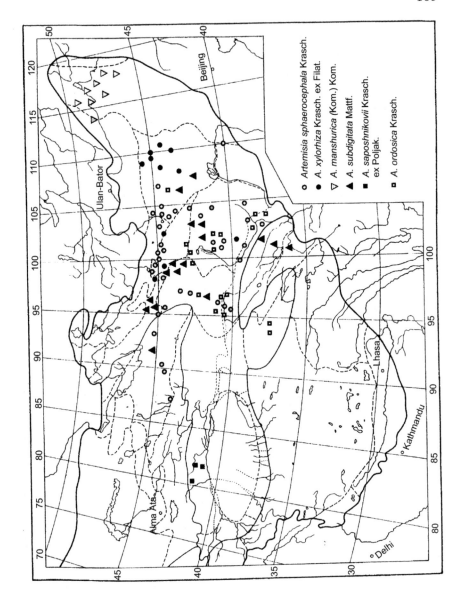

Map 7

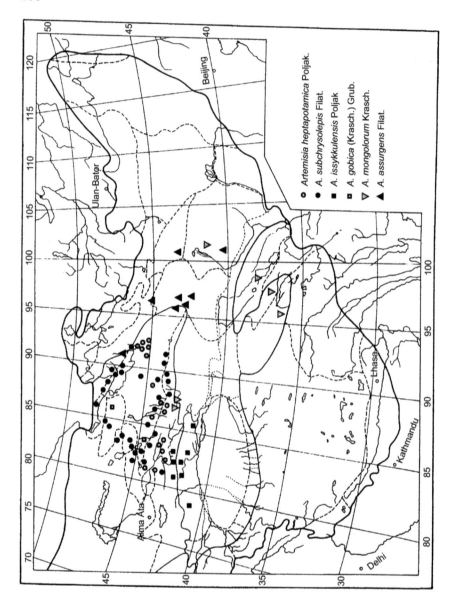

Map 8

INDEX OF LATIN NAMES OF PLANTS

INDEX OF PLANT DISTRIBUTION RANGES

Kaschgaria brachanthemoides (Winkl.) Poljak   4
    komarovii (Krasch. et Rubtz.) Poljak   4
Poljakovia falcatolobata (Krasch.) Grub. et Filat.   2
    kaschgarica (Krasch.) Grub. et Filat   2
Pyrethrum pyrethroides (Kar. et Kir.) B. Fedtsch. ex Krasch   1
Stilpnolepis centiflora (Maxim.) Krasch   4
Tripleurospermum ambiguum (Ledeb.) Franch. et Sav.   1
Waldheimia tridactylites Kar. et Kir.   2

INDEX OF PLANT ILLUSTRATIONS